中国建筑工业出版社
学术著作出版基金项目

CHENG TAINING 程泰宁
ARCHITECTURE WORKS 2015-2021　建筑作品选　2015-2021

程泰宁　著

中国建筑工业出版社

程泰宁

中国工程院 院士
中国建筑设计大师
教授,博士生导师
东南大学建筑设计与理论研究中心主任
筑境设计主持人

CHENG TAINING

Academician of Chinese Academy of Engineering

China Architecture Design Master

Professor, Doctorial Supervisor

Director of Architectural Design & Theory Research Center of Southeast University

Chairman,CCTN Design

立足此时　　立足此地　　立足自己
BASED ON NOW　　BASED ON HERE　　BASED ON MYSELF

有所思

淡淡长空鸟迹遥，蒙蒙烟雨钓情饶。跳出三界不许路，化入无边无垠气。精鹜八极下四字，若无其事人道之眉间。唤醒一风雨蓦而，空对青天话沧桑。

庚寅

诗一首（代序）

凌波学海风拍浪，极目寰宇剑倚天。
跳出三界觅新路，化入自然境为先。

精骛八极开心宇[1]，学究天人追前贤[2]。
唤醒东风化春雨，笑对青山续新篇[3]。

解读

　　建筑为一学海、一江湖，虽说不上波涛汹涌，但明潮暗流从未停息。凌波其上，踏浪而行，很自励，也很自在；凝目这浩渺无垠的宇宙，一人一剑站立在一片岩石、一叶扁舟上，洪荒八极，千山万仞，尽收眼底。畅极，快哉；"三界"：古今、中外、人我是也。只有跳出这三者的羁绊，才能自由自在地走一条自己的探索之路，持正出奇，以新代雄[4]；

　　新在何处？路有千条，就我而言，化入自然即是。技巧、创意与境界的无间融合，浑然天成、自然生成，即是新意；

　　格局、情志，为做好建筑创作之根本；"虽千万人，吾往矣"的气势，亦为一个有责任感的建筑师应有之品格，总揽中外古今文化发展的规律，坚定而自信地走在探索的路上，爽！

　　西风东渐已逾百年，反思彼我之优劣短长，兼收并蓄，并知扬弃，方能闯出新路。东风化雨、润物无声，独立思考、移"风"易"俗"[5]，此乃我与同道之所愿也。

注释

注1：陆机《文赋》："精骛八极，心游万仞。观古今于须臾，抚四海于一瞬。"
注2：司马迁《报任安书》："究天人之际，通古今之变，成一家之言。"
注3：辛弃疾《贺新郎》："我见青山多妩媚，料青山见我应如是。"
注4：萧子显《南齐书·文学传论》："若无新变，不能代雄。"
注5：《奴俗》，清初学者傅山（青主）批评那些拜倒在古人面前之行径。此亦适用于今日之崇洋媚洋之风。"奴"而且"俗"，细思极为到位。

A POEM (AS PREFACE)

Riding on the waves of knowledge, turbulent underneath.
Staring into untold universe, a sword leaned.
Unblocking three realms, a new road explored.
Sublimating into being, intellectual state for the first.

Pursuing no limit, delighted to be[1].
Investigate the One, Chasing sages[2].
Awaking East wind, blowing into spring rain.
Smiling to the past, a new chapter comes[3].

INTERPRETATION

If we regard ARCHITECTURE as an academia sea, a fate river, it's not really with roaring waves, but always turbulent both overtly and covertly. Surfing on which, I feel self-exited and unrestrained all the time.

Staring into the untold universe, standing alone with a sword on a single rock or a tiny boat, with all the living and things, chaos and silence of the whole world in my eyes, I feel cheerful and fulfilled.

Three realms, which means the past and the present, the western and the eastern, the others and the self. Only if we got rid of the block of these three realms, could we be able to explore a road of our own, which benefit from the old but lead to the new. Only innovation lasts in generations[4].

How to innovate? There are thousands of ways. But for me, the answer is to sublimate the being into what it is. The whole integration of technique, concept and intellectual state which makes the building as a natural creation with naturally accomplished unity, is just the innovation.

Artistic conception and intellectual state are foundation of architectural creation. The belief to go against the prevailing trend, is also the essential character of a responsible architect. Understanding the law of development from the past to the present of both Chinese and foreign culture, we are on our own road of innovation. What a satisfaction!

It has been more than a hundred years since the east learning from the west. Reflecting on the strengths and weaknesses between the other and ourselves, and learning from each other, we could be able to find the innovation road. What we want is just the east culture premotion, the independent thinking, the traditional transformation of the architecture[5].

NOTES

1. Quoted from 'Then, the spirit at full gallop reaches the eight limits of the cosmos, and the mind, self-bouyant, will ever soar to new insurmountable heights… Eternity he sees in a twinkling, and the whole world he view in one glance', from Essay on Literature (文赋) written by Lu Ji.
2. Quoted from 'to explore the relationship between the Way of Heaven and the Way of Man, have a thorough understanding of the course of historical development and the changes involved therein, expound my own opinions of the events of the past and present my own system of analysis', from Letter to Ren'an (报任安书), written by Sima Qian.
3. Quoted from 'How charming I see the green mountains, I expect the green mountains regard me the same', from He Xin Lang (贺新郎), written by Xin Qiji.
4. Quoted from 'Without innovation, nothing lasts in generation', from History of Nanqi Dynasty: about Literature (南齐书·文学传论), written by Xiao Zixian.
5. "Nu su": slaved by the vulgar, was firstly mentioned in Fu Shan's critic of those who worship the ancient to the most. It is also appropriately used to describe the kind of xenophilia nowadays.

前 言

一、从1997年"程泰宁建筑作品选"出版算起,这本"作品集"已是第六本了。本来,在网络如此发达的今天,出纸质书籍显得有点不合时宜,但是,24年来,每4～5年出一本已经形成了一个系列,中断了似乎有点可惜。所以还是想继续出版。说到这里,首先要感谢中国建筑工业出版社一直以来的支持,对出版社而言,经济效益是谈不上了,也就是一种情怀在支持吧。

二、从2015年到2021年这六年来,设计项目共38个,但建成的只有15个,此中缘由难以说清,这大概也是目前不少建筑师的共同境遇,郁闷。

三、这两三年,连续参加了几个大型项目国际招标,胜负无论,颇有收获,但更有所触动。除已将这几个项目投标方案列入本书外,有些想法准备留待另书讨论。

四、在第五册与本册之间的2018年,澳大利亚视觉出版集团(IMAGES)出版了大师系列"程泰宁建筑作品选",其中载入已建成的黄岩博物馆与苏步青纪念馆,为避免重复,不再列入本册。

五、真诚感谢为这本书出版做出努力的同事以及我的学生们。

程泰宁

2021年8月 杭州

INTRODUCTION

I. Since the publication of the first Cheng Taining Architecture Works in 1997, this book has been the sixth of the series. Although it seems a little inappropriate to publish paper books in this Info Age, I still want to continue the series as a succession of the 24-year tradition--adding a new volume to the collection every 4-5 years. Here, I would like to thank the CAPB for their continuous support ever since the beginning of my career. There must be the spirit shared by us beyond economic concerns that helped to accomplish the publication of this book.

II. In the past 6 years from 2015 to 2021, we have designed 38 projects, but only 15 of which have been built . The reason for this is hard to explain, but it is also the depressed common situation that most of the architects are faced with.

III. In the past 2-3 years, we have participated in several large-scale international bidding projects in a row, regardless of the outcome, quite rewarding, but even more touching. In addition to include these projects in this book, I prepare to discuss some related thoughts in other books.

IV. In 2018, between the fifth volume and this volume, the IMAGES in Australia have published *Master Series Cheng Taining Architecture Works*，among which Huangyan Museum and the Su Buqing Memorial Hall are not repeated in this volume.

V. Credits go to my colleagues and students for their extraordinary efforts during editing and publication. Thanks a lot!

Cheng Taining
August, 2021, in Hangzhou

CONTENTS 目录

A POEM(AS PREFACE)	6	诗一首（代序）
INTRODUCTION	8	前言
DESIGN WORKS	13	设计作品选

NANJING ART MUSEUM	14	南京美术馆
WENLING MUSEUM	52	温岭博物馆
CHANGCHUN WORLD SCULPTURE & ART MUSEUM	70	长春世界雕塑艺术博物馆
JUNKANG FINANCIAL PLAZA	84	上海君康金融广场
THE MUSEUM AND READING CENTER IN NEW CAMPUS OF XI'AN JIAOTONG UNIVERSITY	104	西安交通大学新校区博物馆及多功能阅览中心
LONGYAN MUNICIPLE PARTY SCHOOL	126	龙岩市委党校
QINGDAO (RED ISLAND) RAILWAY STATION	138	青岛（红岛）铁路客站
SUZHOU HIGH SCHOOL-SUZHOU BAY CAMPUS	158	苏州中学苏州湾校区
CANGQIAN CAMPUS OF HANGZHOU NORMAL UNIVERSITY	168	杭州师范大学仓前校区
HANGZHOU WEST RAILWAY STATION COMPLEX	184	杭州西站及城市综合体
CONCEPTUAL URBAN DESIGN PROPOSAL FOR THE OUTDOOR STADIUM, INDOOR GYMNASIUM AND NATATORIUM AND NEW CONVENTION CENTER IN XIAMEN	202	厦门一场两馆、新会展中心城市设计及建筑概念设计方案
THE NEW CONVENTION CENTER OF XIAMEN	210	厦门新会展中心
INTERNATIONAL COMPETITION PROPOSAL OF URBAN DESIGN AND PHASE II ARCHITECTURE DESIGN OF NEW BEIJING INTERNATIONAL EXHIBITION CENTER	226	北京新国展二、三期城市设计及建筑概念设计方案
GUOSHEN MUSEUM PROPOSAL	244	国深博物馆建筑设计方案
GRAND THEATRE PROPOSAL OF BEIJING CITY SUB-CENTER	254	北京城市副中心大剧院建筑设计方案
CHINESE NATIONAL MUSIC CENTER,JIANGYIN	268	中华国乐中心·江阴
NANJING YUHUA HUAYI HOTEL	278	南京雨花华邑酒店
DOWN HOLE HOTEL OF XUZHOU GARDEN EXPO	290	徐州园博会宕口酒店
CHINA WATER CONSERVANCY ENGINEERING SCIENCE AND TECHNOLOGY MUSEUM	304	中国水工科技馆

LANGUAGE, ARTISTIC CONCEPTION AND INTELLECTUAL STATE AN INTERview WITH CHENG TAINING ON HIS ARCHITECTURAL THOUGHTS	316	语言·意境·境界 ——程泰宁院士建筑思想访谈录
CHRONOLOGICAL LIST OF PROJECTS (2015-2021)	344	作品年表（2015-2021）
POSTSCRIPT	347	后记

设计作品选　DESIGN WORKS

南京美术馆
NANJING ART MUSEUM

项目地点　中国・江苏・南京
项目规模　97275.9 m²
合 作 者　王大鹏、汪毅、蓝楚雄、柴敬、刘翔华、王岳锋、刘彦伯
项目时间　2016 年设计，国际设计竞赛中标，2021 年竣工

Location　Nanjing, Jiangsu Province, China
Gross floor area　1047069 ft² (97275.9 m²)
Associates　Wang Dapeng, Wang Yi, Lan chuxiong, Chai Jing, Liu Xianghua, Wang yuefeng, Liu yanbo
Status　Designed in 2016, Won the bidding in international competition. Completed in 2021.

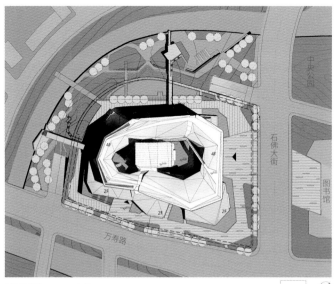

总平面图　General plan

对页：西南方向鸟瞰（陈畅 拍摄）
Opposite: Aerial view from the southwest
(Photographer, Chen Chang)

东北方向鸟瞰（陈畅 拍摄）　Aerial view from the northeast (Photographer, Chen Chang)

南京美术馆位于国家级江北新区商业核心区，北望老山，南眺长江，是江北地区青龙绿带上的一颗璀璨明珠。

打破传统"艺术殿堂"的高冷形象，强调功能复合，加强美术馆的开放性，是当代文博类建筑的发展趋势。本设计充分考虑这一特点，打造了一个能吸引广大市民的充满活力的公共空间。

设计中，美术馆距基座 18m 架空，最大限度地引入了自然山水与城市景观，建筑成为全方位对外开放的立体园林。这一立体园林与以水墨画为意象的中央大厅的彩釉玻璃外墙，准确地表达了美术馆的艺术形象，更体现了一种"中国调性"。

Nanjing Art Museum is located in the central business district of Nanjing the Jiangbei New Zone, facing Laoshan Mountain in the north and the Yangtze River to the south. The project is positioned like a brilliant pearl in the Qinglong greenbelt of the Jiangbei area (north of the Yangtze River).

This design breaks the traditional mold of a museum as a cold and lofty "palace of art", putting the emphasis on integrated functions, and reinforcing the openness of the building, following contemporary trends in cultural values. The project intends to create a vibrant and lively public space that is capable of attracting massive crowds of locals and tourists alike.

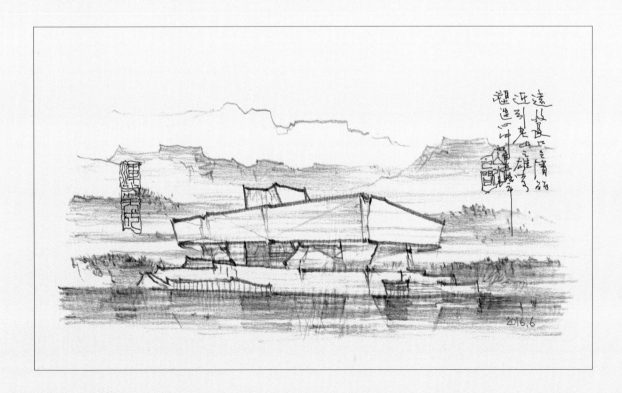

下图：场地区位分析 Below: Site location analysis
对页：功能分析 Opposite: Program analysis

In this design, the museum is elevated 59 feet (18 meters) above the foundations in order to attract the public eye and to blend in more appealingly with the natural scenery and the urban landscape. The building is envisioned as a three-dimensional garden open to the public from all directions. This three-dimensional garden utilizes the image of an ink painting as the basis for the colored glass wall of the central hall, so as to clearly define the artistic image of the museum and to express a "Chinese tonality".

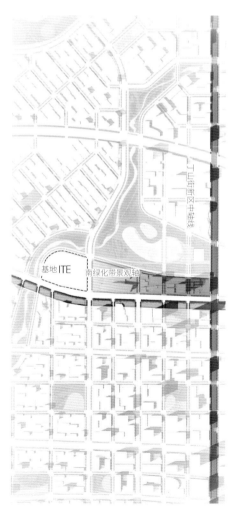

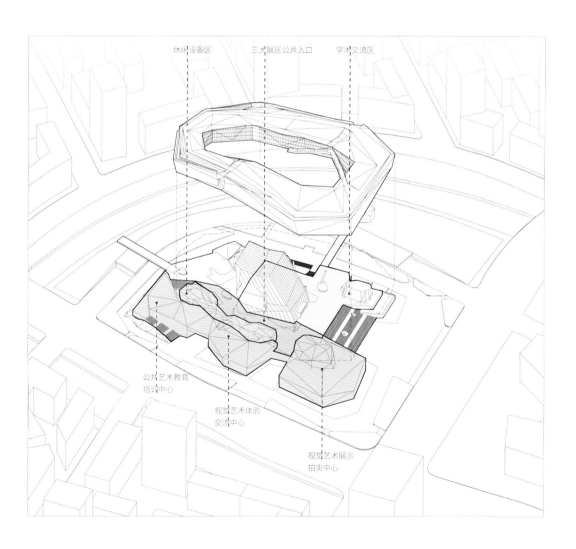

下图：根系金陵传统，面向江北未来，打造中国现代建筑新形象
Below: Rooted in Nanjing tradition, but facing the future for the northern areas of the Yangtze River, and creating a new image of modern building in China.

对页：向绿地水面开放，与城市中轴连接，融入环境，点亮城市
Opposite: Open to green space and water, linked with urban axis, integrated with environment and enhancing the cityscape.

向绿地水面开发
与城市中轴连文
融入环境
点亮城市

公共开放，立体园林 (陈畅 拍摄)
Open to the public, three-dimensional garden (Photographer, Chen Chang)

1　中央大厅　　　　　　　　Central hall
2　美术馆　　　　　　　　　Art museum
3　画家工作室及办公室　　　Painter's studio and office
4　视觉艺术展示拍卖中心　　Visual art exhibition and auction center
5　视觉艺术体验交流中心　　Visual art experience and exchange center
6　公共艺术教育培训中心　　Public art education and training center
7　二层开放共享平台　　　　Second-floor shared platform

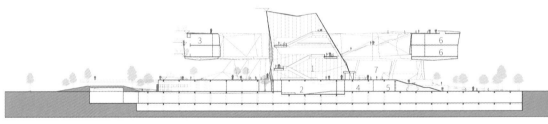

剖面图　Section

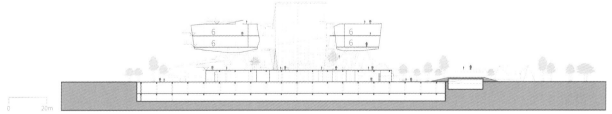

剖面图 Section

1	门厅	Entrance hall
2	教育培训中心	Education and training center
3	商店	Retailis
4	视觉艺术体验交流中心	Visual art experience and communication center
5	展厅	Exhibition hall
6	视觉艺术展示拍卖中心	Visual art exhibition and auction center
7	拍卖厅	Auction hall
8	报告厅	Lecture hall
9	会议室	Meeting room
10	门厅	Lobby
11	餐厅	Canteen
12	教室	Classroom
13	库房	Storeroom
14	下沉广场	Sunken plaza

1	主入口	Main entrance
2	中央大厅	Central hall
3	视觉艺术展示拍卖中心	Hall of visual center
4	办公室	Office
5	库房	Storeroom
6	休闲吧	Coffee bar
7	美术馆商业	Retails
8	培训教室	Training classroom
9	共享平台	Public platform
10	下沉广场上空	Upper-level of sunken plaza
11	景观廊桥	Landscape bridge

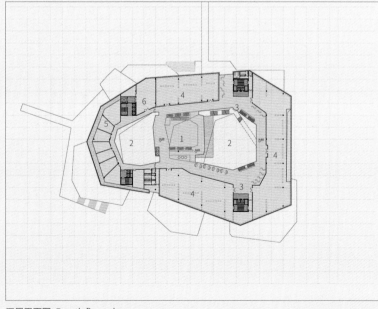

四层平面图 Fourth floor plan

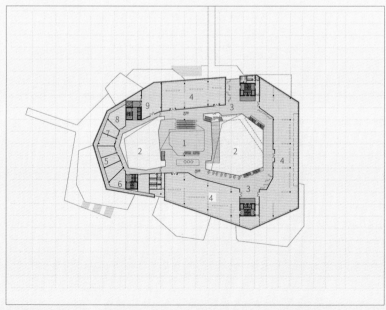

三层平面图 Third floor plan

	中文	English
1	中央大厅上空	Void above of central hall
2	上空	Void
3	公共展廊	Public exhibition corridor
4	展厅	Exhibition hall
5	办公室	Office
6	馆长室	Curator office
7	会议室	Meeting room
8	接待室	Reception room
9	库房	Storeroom

	中文	English
1	中央大厅上空	Void above of central hall
2	上空	Void
3	公共展廊	Public exhibition corridor
4	展厅	Exhibition hall
5	画家工作室	Artists' office
6	库房	Storeroom

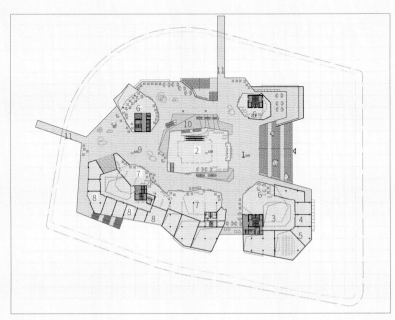

二层平面图 Second floor plan

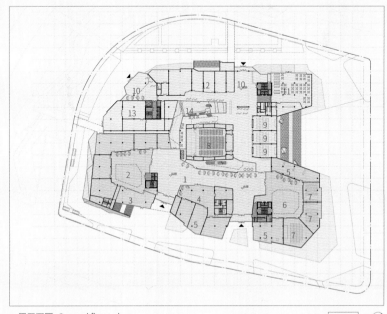

一层平面图 Ground floor plan

下图及对页：东南方向外景（陈畅 拍摄）　Below and opposite: Southeast View（Photographer, Chen Chang）

下图：西南角外景（陈畅 拍摄） Below: Southwest view (Photographer, Chen Chang)
对页：建筑局部（陈畅 拍摄） Opposite: Aspects (Photographer, Chen Chang)

下图：办公入口（陈畅 拍摄）　Below: Office entrance （Photographer, Chen Chang）
对页：西北方向外景（陈畅 拍摄）　Opposite: Northwest view （Photographer, Chen Chang）

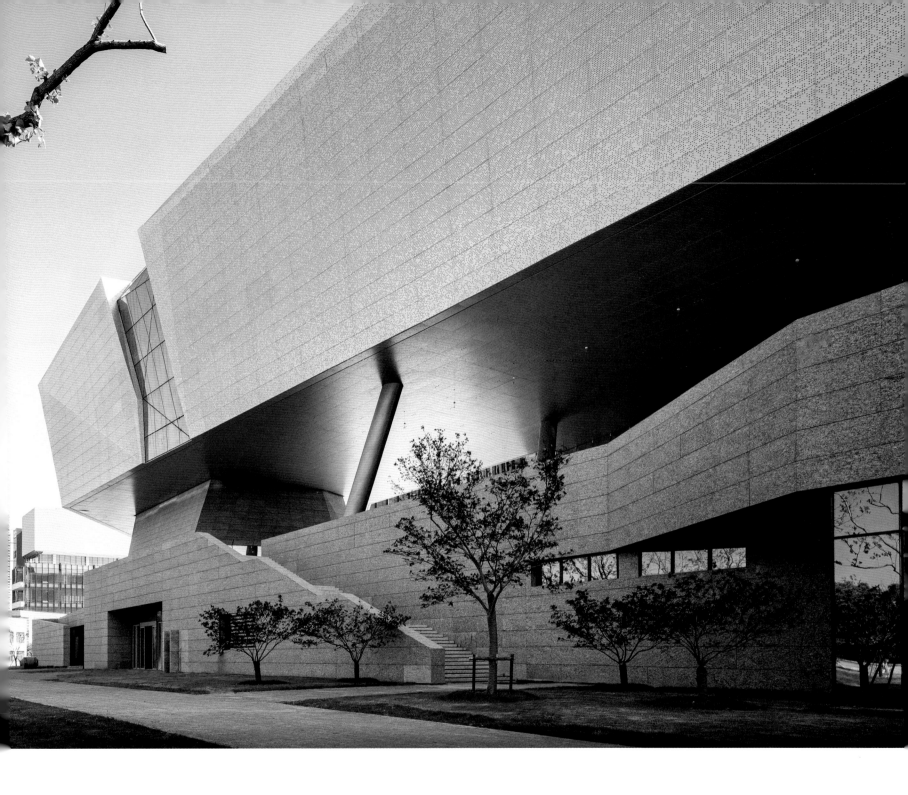

左上：办公入口（陈畅 拍摄）
Left upper: Office entrance（Photographer, Chen Chang）
左下：东北方向局部外景（陈畅 拍摄）
Left lower: Northeast view（Photographer, Chen Chang）
对页：主入口东北方向外景（陈畅 拍摄）
Opposite: Northeast view of the main entrance
（Photographer, Chen Chang）

下图：西北方向局部外景（陈畅 拍摄） Below: Northwest view （Photographer, Chen Chang）
对页：主入口礼仪广场（陈畅 拍摄） Opposite: Etiquette square at the main entrance （Photographer, Chen Chang）

下左上：主入口礼仪台阶（陈畅 拍摄）Bottom left upper: Ritual steps of main entrance （Photographer, Chen Chang）
下右上：西南方向内环透视（陈畅 拍摄）Bottom right upper: Circular perspective from southwest （Photographer, Chen Chang）
下左下：公共平台局部（陈畅 拍摄）Bottom right below: Outdoor platform （Photographer, Chen Chang）
下右下：东北内环透视（陈畅 拍摄）Bottom left below: Circular perspective from southeast （Photographer, Chen Chang）
对页："内环"局部透视（陈畅 拍摄）Opposite: The "Inner ring" perspective （Photographer, Chen Chang）

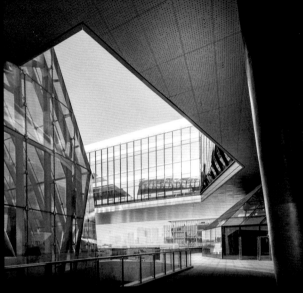

下图：公共开放平台（陈畅 拍摄）Below: Public platform （Photographer, Chen Chang）
对页：室外开放展区（陈畅 拍摄）Opposite: Public exhibition area （Photographer, Chen Chang）

散落布置在高架平台及下沉小广场中的文化休闲与商业服务设施，为观众和市民在此停留休憩创造了条件。功能复合，使美术馆具有更大的吸引力。

Extensive facilities for cultural enjoyment and commercial services, strewn over the elevated platform and recessed in the small sunken square, create a convenient place for visitors and citizens to stop and relax. These integrated functions have endowed the art museum with greater qualities of attraction.

上图：夕阳剧场（陈畅 拍摄）Top: Sunset theater (Photographer, Chen Chang)
下图及对页："绿谷"（陈畅 拍摄）Below and opposite: Green valley (Photographer, Chen Chang)

下图及对页：中央礼仪大厅（陈畅 拍摄）
Below and opposite: Central etiquette hall (Photographer, Chen Chang)

下图及对页：中央礼仪大厅（陈畅 拍摄） Below and opposite: Central etiquette hall （Photographer, Chen Chang）

下图及对页：展示拍卖中心室内（陈畅 拍摄） Below and opposite: Exhibition and auction center (Photographer, Chen Chang)

下左图：学术报告厅（陈畅 拍摄）**Left below : Lecture hall** (Photographer, Chen Chang)
下右上及下右下图：展厅（陈畅 拍摄）**Right top and right bottom: Exhibition hall** (Photographer, Chen Chang)
对页：室内空间（陈畅 拍摄）**Opposite: Indoor space** (Photographer, Chen Chang)

温岭博物馆
WENLING MUSEUM

项目地点　中国·浙江·台州
项目规模　9450 m²
合 作 者　陈玲、刘翔华、史晟、王忠杰
项目时间　2011 年设计，2018 年竣工

Location　Taizhou, Zhejiang Province, China
Gross floor area　101718.95 ft² (9450 m²)
Associates　Chen Ling, Liu Xianghua, Shi Sheng, Wang Zhongjie
Status　Designed in 2011. Completed in 2018.

设计中，我们探索了数字化时代非线性建筑的适宜性运用。

首先，设计适宜于地域文化的表达。"石文化"是温岭市四大文化中排名第一的重要地域文化，形如山石的建筑造型是对当地文化的回应。

其次，设计适宜于场地环境。项目场地位于石夫人山脚下，周围高楼林立，用地狭小呈三角形。在这样的场地条件下，通常的建筑造型极易湮没在水泥森林之中，成为高楼的裙房。而设计将建筑的功能与造型整合，如同石夫人山下一块散落的山石，不仅与自然环境相融，也能将这一重要的公共建筑从城市环境中凸显出来。

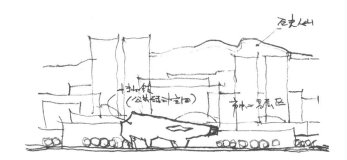

对页：鸟瞰——非线性建筑的中国调性（陈畅 拍摄）
Opposite: Aerial view of non-linear architecture with a Chinese tonality (Photographer, Chen Chang)
下页：沿河透视图（陈畅 拍摄）
Next page: Riverside view (Photographer, Chen Chang)

对页：西北向外景（陈畅 拍摄）Opposite: Northwest view (Photographer, Chen Chang)

1　设备储藏室　　Equipment storeroom
2　临时展厅　　　Temporary exhibition hall
3　展厅　　　　　Exhibition hall
4　走廊　　　　　Corridor
5　展厅　　　　　Exhibition hall
6　休息区　　　　Lounge
7　多功能厅　　　Multifunction hall

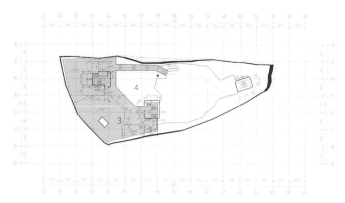

三层平面图　Third floor plan

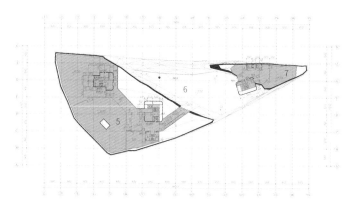

二层平面图　Second floor plan

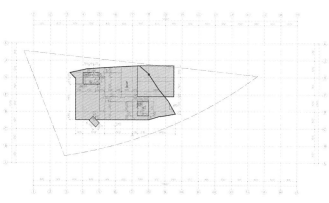

一层平面图　Ground floor plan

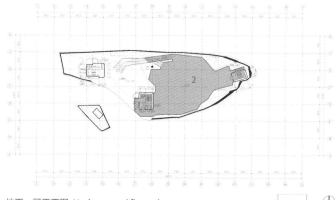

地下一层平面图　Underground floor plan

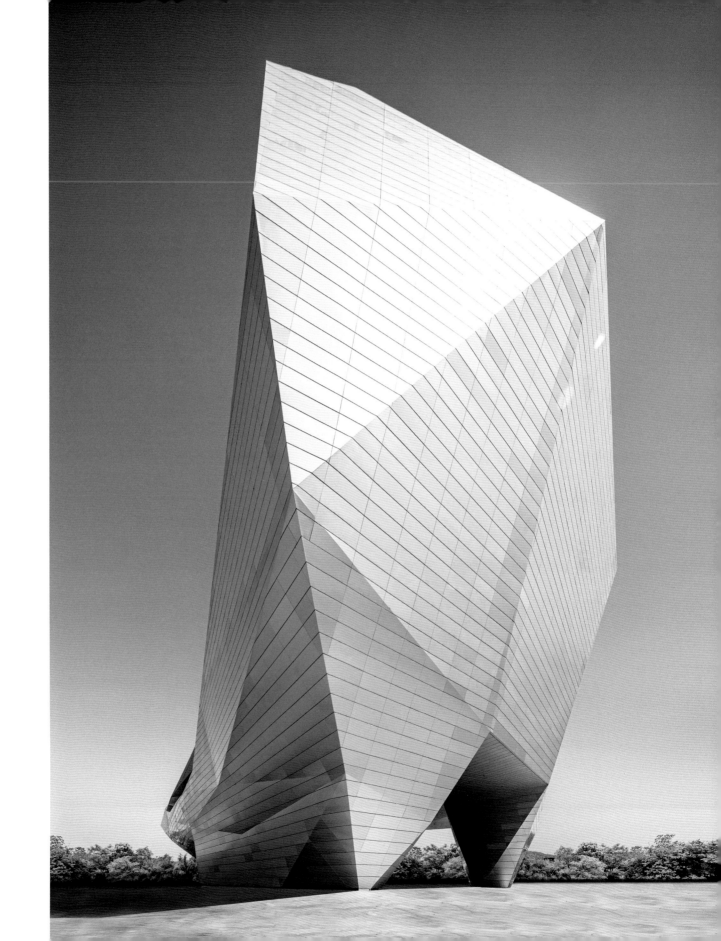

对页：东北向鸟瞰（陈畅 拍摄）Opposite: Northwest aerial view (Photographer, Chen Chang)

In this project, the designers explored the application of non-linear architecture in the digital age.

Firstly, the design suits the need of expressing regional culture. The "stone culture" was the earliest significant regional culture among Wenling city's four major cultures, with which the stone-resembling architectural form aims to resonate.

Secondly, the design should fit the site environment. Located at the foot of the Stone Lady Mountain, the narrow triangular project site and is surrounded by high buildings. Under such circumstances, any usual building forms would seem buried in the surrounding concrete jungle and become podium of the skyscapers. However, the project design integrates the functions of the museum into the form--like a mountain rock scattered from the Stone Lady Mountain. Thus, it does not just fuse into the natural environment but also highlights this important public building within the urban context.

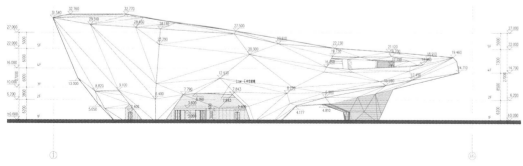

北立面图 North elevation

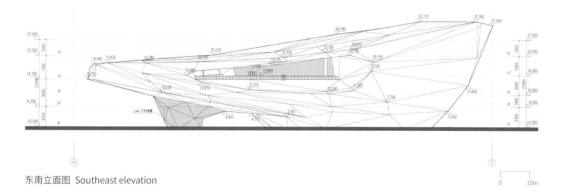

东南立面图 Southeast elevation

下图：设计草图 Below: Design sketch
对页：东北角外景（陈畅 拍摄）Opposite: Northwest view（Photographer, Chen Chang）

下图：建筑局部（陈畅 拍摄）Below: Aspects（Photographer, Chen Chang）
对页：西南方向外景（陈畅 拍摄）Opposite: Southwest view（Photographer, Chen Chang）

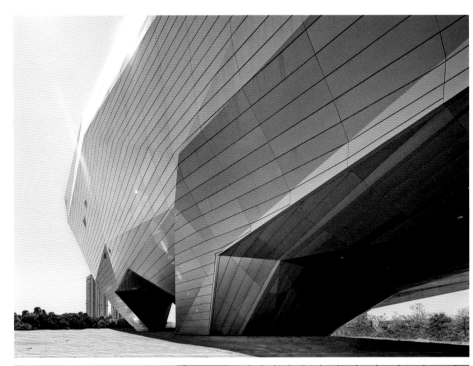

参数化设计的运用,为我们表达非线性建筑造型提供了更为多样的手段,大大拓展了建筑的艺术表现力。

The use of parametric modeling design helped to create a more diversified means of expression for the modeling of non-linear architecture, which greatly empowered the artistic strength of the architecture.

下图：建筑局部（陈畅 拍摄）Below: Aspects（Photographer, Chen Chang）
对页：建筑底层架空营造室外空间（陈畅 拍摄）Opposite: Elevating ground floor to create outdoor space（Photographer, Chen Chang）

下图：建筑局部（陈畅 拍摄） Below: Aspects (Photographer, Chen Chang)
对页：入口局部（陈畅 拍摄） Opposite: Entrance view (Photographer, Chen Chang)

上图：前厅顶部（陈畅 拍摄） Top: Ceiling of lobby (Photographer, Chen Chang)
下图：室内——门厅入口（陈畅 拍摄） Below: Interior—entrance of lobby (Photographer, Chen Chang)
对页：前厅（陈畅 拍摄） Opposite: Lobby (Photographer, Chen Chang)

长春世界雕塑艺术博物馆
CHANGCHUN WORLD SCULPTURE & ART MUSEUM

项目地点　中国·吉林·长春
项目规模　18000 m²
合 作 者　王大鹏、吴旭斌、汪毅、刘鹤群、吕思扬、谢潘扬
项目时间　2014 年设计，2018 年竣工

Location　Changchun, Jilin Province, China
Gross floor area　193750.4 ft² (18000 m²)
Associates　Wang Dapeng, Wu Xubin, Wang Yi, Liu Hequn, Lv Siyang, Xie Panyang
Status　Designed in 2014. Completed in 2018.

　　长春世界雕塑艺术博物馆以展示雕塑特色文化、传播知识、陶冶性情为目的，同时为雕塑的收藏、保护、展示、研究提供现代化的场所和窗口。

The construction purpose of Changchun World Sculpture & Art Museum is both to display the unique culture of sculpture, and to disseminate the knowledge and cultivate the temperament of sculpture. Meanwhile, it provides a modern place for the collection, protection, exhibition and research of sculpture.

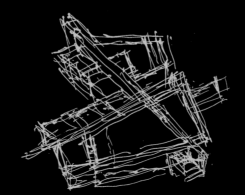

对页：建筑外景（黄临海 拍摄）
Opposite: Exterior perspective
(Photographer, Huang Linhai)

入口透视图（陈畅 拍摄）Entrance perspective（Photographer, Chen Chang）

　　场地位于长春国际雕塑公园东南角，优越的周边自然环境使我们思考如何让建筑自然和谐地处于雕塑公园中，并且既能体现出文化建筑自身的特点又与雕塑公园的整体气质协调，这是我们设计的出发点。

　　设计理念以"经天纬地、雕刻时光"为出发点。本方案整体以经纬纵横的两条时光长廊为构架，其中南北方向的长廊与整个公园的核心雕塑形成对位关系；东西方向的长廊串联起了门厅、中央大厅、展厅以及研究办公等空间序列。整个建筑如同巨石破土而出，体积感强烈厚重，极具雕塑感，虽出人工，宛如天成。建筑利用天窗和立面的缝隙对不同朝向的光线进行捕捉处理，影随光移，空间被时光雕刻，并且为雕塑的展陈营造了良好的氛围。

The site is located in the southeast corner of Changchun International Sculpture Park. As a starting point of the proposal, the superior surrounding natural environment leaves a question to architect about how to make a building in the sculpture park in a harmonious way, which can not only reflect the characteristics of cultural architecture but also coordinate with the overall atmosphere of the sculpture park.

The design takes "heaven and earth, carving time" as the concept. The general layout is constructed with two time-corridors of latitude and longitude, among which the north-south corridor forms a relationship in accordance with the core sculpture of the whole park. The east-west corridor connects the lobby, the central hall, the exhibition hall and the research office. The whole building is like a huge stone breaking out of the earth, with a strong sense of volume, massiness and sculpture-like. The building uses skylights and glass gaps on facade to capture the light in different directions. The shadows move with the light, while the space is carved by time, and a good atmosphere is created for the exhibition of the sculpture.

对页：正立面外景（黄临海 拍摄）Opposite: Main facade view（Photographer, Huang Linhai）

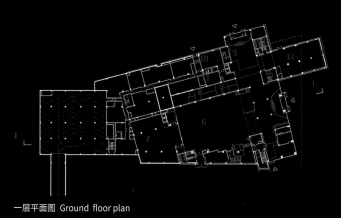

1	主门厅	Entrance hall
2	问询票务	Inquiries/Tickets
3	纪念品展卖	Souvenir
4	冷餐间	Buffet space
5	雕塑台	Sculpture platform
6	礼仪大厅(中庭)	Central hall
7	展厅	Exhibition hall
8	展廊	Exhibition corridor
9	次门厅	Secondary entrance hall
10	贵宾接待室	VIP lounge
11	化妆间	Dressing room
12	咖啡吧	Cafe bar

一层平面图 Ground floor plan

1	主门厅上空	Void above entrance hall
2	雕塑台	Sculpture platform
3	礼仪大厅上空	Void above central hall
4	展厅上空	Void above exhibition hall
5	办公门厅	Lobby of office space
6	工作室	Studios
7	会议室	Meeting room
8	展厅	Exhibition hall
9	展廊	Exhibition corridor
10	次门厅上空	Void above secondary entrance
11	不上人屋面	Roof

二层平面图 Second floor plan

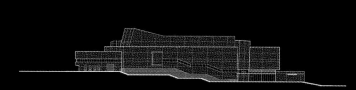

南立面图 South elevation

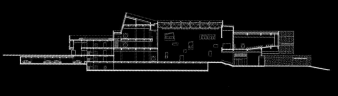

剖面图 Section

总平面图　General plan

下图：北立面图（黄临海 拍摄）Below: North elevation（Photographer, Huang Linhai）
对页：局部透视（黄临海 拍摄）Opposite: Aspect view（Photographer, Huang Linhai）

下图及对页：展厅内景（黄临海 拍摄） Below and opposite: Interior of the exhibition hall (Photographer, Huang Linhai)

下图：展厅内景（黄临海 拍摄）Below: Interior of the exhibition hall （Photographer, Huang Linhai）
对页：室内通廊（黄临海 拍摄）Opposite: Interior passage （Photographer, Huang Linhai）

上海君康金融广场
JUNKANG FINANCIAL PLAZA

项目地点　中国 · 上海
项目规模　105000 m²
合作设计　殷建栋、吴妮娜、杨涛、刘翔华、朱文婧、袁越、陈鑫、王政、
　　　　　　古振强、周逸
项目时间　2013 年设计，2020 年竣工

Location　Shanghai, China
Gross floor area　1130211 ft² (105000 m²)
Associates　Yin Jiandong, Wu Nina, Yang Tao, Liu Xianghua, Zhu Wenjing,
　　　　　　　Yuan Yue, Chen Xin, Wang Zheng, Gu Zhenqiang, Zhou Yi
Status　Designed in 2013. Completed in 2020.

　　项目位于上海市浦东新区后滩板块核心区域，世博园区南侧。建筑包括 5 栋由空中连廊连接的办公楼，并配有文化及商业功能。

Located south of the World Expo Garden at the core area of Houtan in Pudong New District, Shanghai, Junkang Financial Plaza consists of five office buildings, which are connected to each other by air corridors, and it serves both cultural and commercial functions.

对页：沿耀元路外景（陈畅 拍摄）
Opposite: View along the Yaoyuan Street （Photographer, Chen Chang）

对页：夜景（姚力 拍摄） Opposite: View at night (Photographer, Yao Li)

  设计以"海上花"为创意，诠释了企业文化，中国传统文化中的理想，同时，以层向错位的手法，打造了一个富有张力和雕塑感的建筑形象，成为沿江地标性建筑。

  同时，通过庭院、连廊、底层商业与沿街绿化的设置，我们希望实现办公空间、景观空间、商业空间、城市空间相互渗透交融，打造一个绿色生态的立体城市山水园林。

  建筑细部处理细微、精致，以较高的设计完成度区别一般建筑。

With the design modeled on the creative concept of "Flowers of Shanghai", this structure embodies the ideal qualities of Chinese traditional culture as a method of interpreting corporate culture.

The plaza has become a landmark of riverfront architecture. Through the configuration of the courtyards, connected corridors, ground-floor business, and street landscape, the design is intended to realize the interpenetration and fusion of office space, landscaped space, commercial space, and urban space for the creation of an organic, three-dimensional urban landscape garden.

The details of the building are subtle and exquisite. Therefore, the building is different from other buildings with its high design completion.

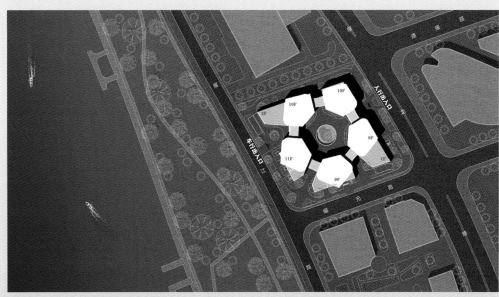

总平面图 General plan

对页：沿江外景（姚力 拍摄）Opposite: View along Huangpu River（Photographer, Yao Li）

1	门厅	Entrance hall	6	健身房	Fitness
2	庭院	Courtyard	7	门厅	Over the entrance hall
3	银行	Bank	8	办公区	Office
4	品牌产品展示厅	Brand products showroom	9	连廊	Corridor
5	咖啡简餐	Café	10	餐饮区	Catering

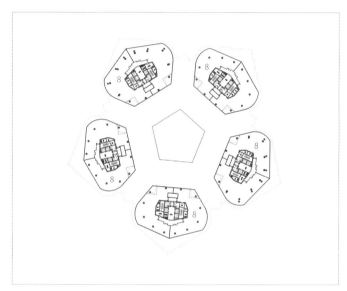

四层平面图 Fourth floor plan

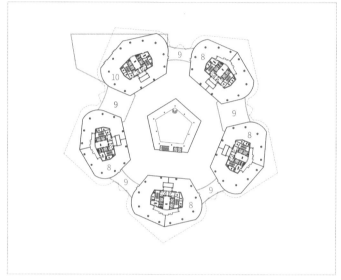

三层平面图 Third floor plan

二层平面图 Second floor plan

一层平面图 Ground floor plan

对页：沿耀元路外景（陈畅 拍摄）Opposite: View along the Yaoyuan Street（Photographer, Chen Chang）

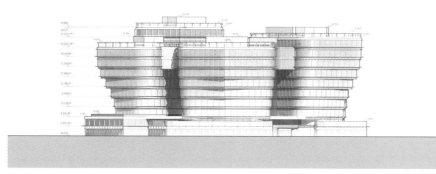

东立面图 East elevation

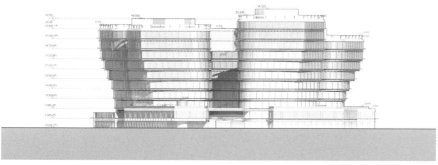

西立面图 West elevation

下图：局部（姚力 摄影）Below: Aspect (Photographer, Yao Li)
对页：建筑外景（姚力 摄影）Opposite: Exterior perspective (Photographer, Yao Li)

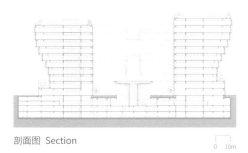

剖面图 Section　0　10m

下图及对页：局部（陈畅 摄影） Below and opposite: Aspect （Photographer, Chen Chang）

下图：设计细部 Below: Detailed design

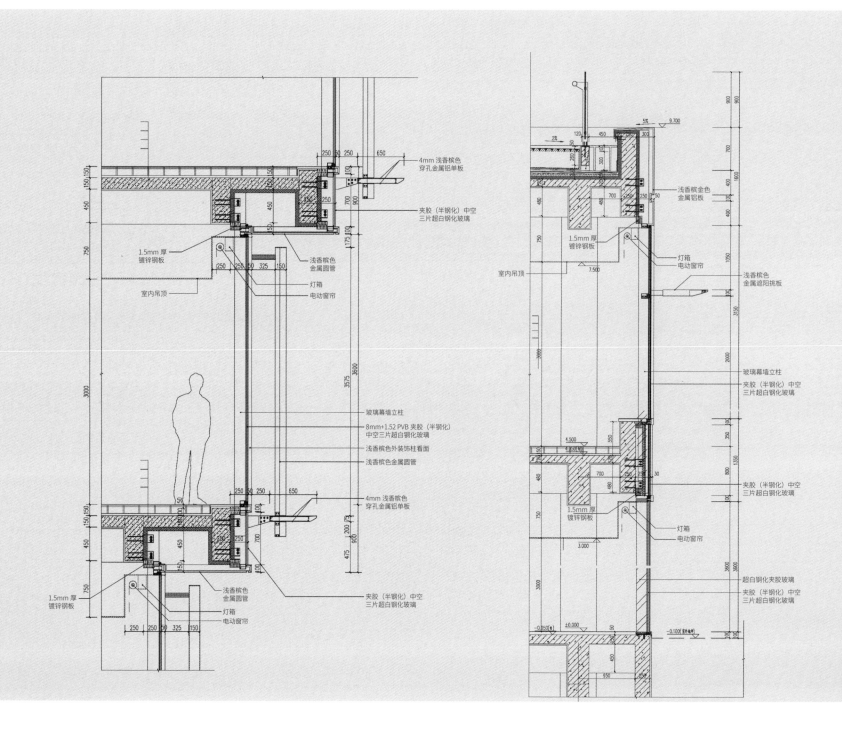

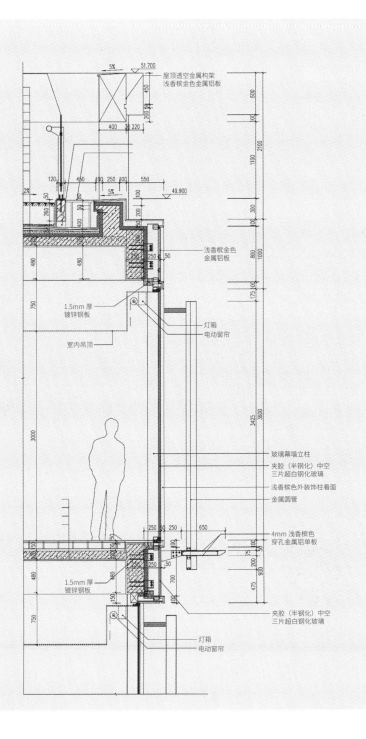
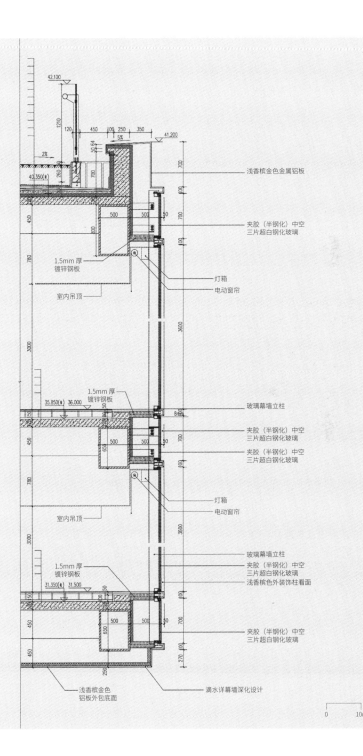

下图：下沉广场（陈畅 拍摄） Below: Sunken plaza (Photographer, Chen Chang)
对页：局部（姚力 拍摄） Opposite: Aspect (Photographer, Yao Li)

下图：从庭院看建筑（姚力 拍摄）Below: View from courtyard（Photographer, Yao Li）
对页：下沉广场（姚力 拍摄）Opposite: Sunken plaza（Photographer, Yao Li）

下图：鸟瞰（姚力 摄影）Opposite: Aerial view（Photographer, Yao Li）
对页：沿世博大道外景（姚力 拍摄）Below: View along the Expo Boulevard（Photographer, Yao Li）

西安交通大学新校区博物馆及多功能阅览中心
THE MUSEUM AND READING CENTER IN NEW CAMPUS OF XI'AN JIAOTONG UNIVERSITY

项目地点　中国·陕西·西安
项目规模　35566 m²
合 作 者　王大鹏、蓝楚雄、杨涛、刘翔华、邱培昕、王静辉、盛思源、
　　　　　裘梦颖、冯单单
项目时间　2017年设计，2020年竣工

Location　Xi'an, Shaanxi Province, China
Gross floor area　382829.24 ft² (35566 m²)
Associates　Wang Dapeng, Lan Chuxiong, Yang Tao, Liu Xianghua,
　　　　　　　Qiu Peixin, Wang Jinghui, Sheng Siyuan, Qiu Mengying,
　　　　　　　Feng Dandan
Status　Designed in 2017. Completed in 2020.

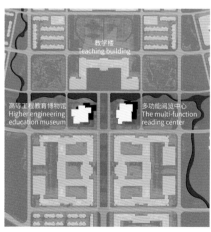

总平面图　General plan

对页：全景鸟瞰（叁山影像　拍摄）
Opposite: Aerial view
(Photographer, Trimont Image)

对页：设计草图 Opposite: Design sketch

项目位于西安交通大学科技创新港科创基地核心位置，处于校园中轴与活力廊道的交汇处，与主楼呈"品"字形布局。8号楼是学校的工程博物馆，9号楼为多功能阅览中心。设计的创意来自于中国四大发明之活字印刷，利用简洁纯粹的方形体量组合，形成错落有致的形体关系，如同活字印刷的模块，又像是老交大四大发明广场上的雕塑，将新老校区的时空记忆联系起来。各个功能体块之间形成的高耸狭缝空间作为一种特殊的公共空间存在，增加了空间的趣味性和神秘感。现代并充满度感、机械感的造型十分符合西安交大著名工程院校的国际形象。

The project is located at the core of the Science and Technology Innovation Base of Xi'an Jiaotong University, at the intersection of the campus axis and the vitality corridor, and in a tripod shape layout with the main building. Building No.8 is the school's engineering museum; Building No.9 is a multi-functional reading center. The creativity of the design comes from the movable-type printing of the four great inventions of China. It uses the simple and pure square volume combination to form the scattered shape relationship, just like the module of movable-type printing, and also like sculpture in the square of the four great inventions in the old campus of Jiaotong University, which connects the space-time memory of the new and old campus. As a unique public space, the tall slit space formed between each functional block increases the interest and mystery of the space. Modern and full of strength, mechanical sense of the shape is very consistent with the international image of Xi'an Jiaotong University as a famous engineering school.

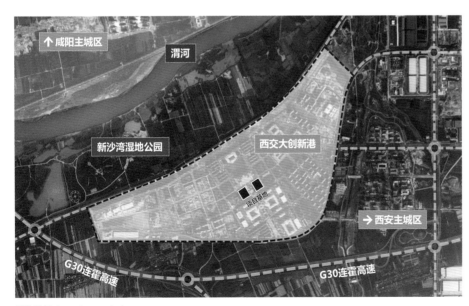

区位分析 Location analysis

对页：夜景鸟瞰（叁山影像 拍摄） Opposite: Aerial view at night (Photographer, Trimont Image)

1	南门厅	South lobby
2	咖啡厅	Cafe
3	多功能厅	Multi-Function hall
4	VIP 接待室	VIP room
5	西门厅	West lobby
6	就餐自习区	Dining & study area
7	庭院	Courtyard
8	备餐	Meal preparation
9	北门厅	North lobby
10	包间	Private room
11	庭院上空	Void above courtyard
12	景观水池	Landscape pond
13	多功能厅上空	Void above multi-function hall
14	自习区	Study area
15	屋面	Roof

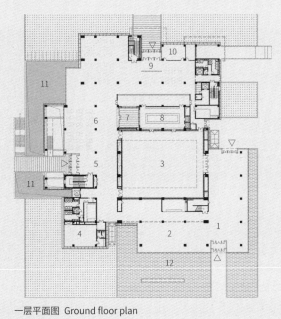

一层平面图 Ground floor plan

二层平面图 Second floor plan　　三层平面图 Third floor plan

0　5m

对页：多功能阅览中心主入口正透视（叁山影像 拍摄）Opposite: Main entrance of multi-function reading center（Photographer, Trimont Image）

多功能阅览中心外立面则采用了"玻璃幕墙＋穿孔铝板"双层幕墙系统。浅灰色的穿孔铝板与综合大楼的淡黄色石材相协调，保证了室内空间具有良好的透光性，又能避免阳光直射，并且还强化了建筑的体积感。穿孔图案则来自对传统清水砖墙肌理的提取与转译，使得现代简约的铝板幕墙具有了一定的文化韵味。浅灰色的半透明穿孔铝板与室内的木饰面板搭配，使得阅览空间光线柔和而富有趣味性，非常契合多功能阅览中心开放共享、多元创新的调性。

The multi-function reading center uses a double-layer curtain wall system of "glass curtain wall + perforated aluminum plate". The light grey of the perforated aluminium panels perform in harmony with the light yellow stone of the complex building. The perforated plates ensure that the interior space has good light, avoiding direct sunlight, and also emphasize the volume of the building. The perforation pattern comes from the extraction and translation of the texture of the traditional water brick wall, which makes the modern and simple aluminum curtain wall have a certain cultural form. The light grey translucent perforated aluminum panels are paired with interior wood panels to make the light of reading space soft and interesting, which perfectly fits the open, sharing and innovative spirit of the multi-functional reading center.

对页：多功能阅览中心夜景（陈畅 拍摄）Opposite: View of multi-function reading center at night (Photographer, Chen Chang)

1	就餐自习区	Dining & Study Area	9 泳池	Swimming Pool
2	下沉庭院	Sunken Garden	10 均衡水池	Balance Pool
3	空调机房	Air-conditioning Control Room	11 水处理机房	Water-cycle Room
4	备餐	Meal Preparation	12 汽车坡道	Car Ramp
5	中央厨房	Central Kitchen	13 洗消中心	Decontamination Center
6	弱电机房	Weak Electricity Engine Room	14 走道	Walkway
7	休息厅	Lounge	15 咖啡厅	Cafe
8	多功能厅	Multi-functional Hall		

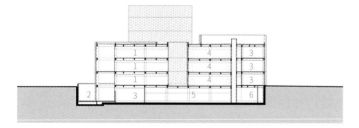

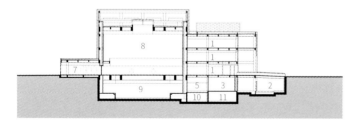

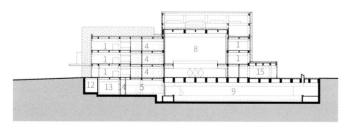

组合剖面图 Sections

下图：多功能阅览中心下沉庭院灰空间（叁山影像 拍摄）Below: Pseudo-public space in sunken garden of multi-function reading center（Photographer, Trimont Image）
对页：多功能阅览中心南立面局部（叁山影像 拍摄）Opposite: South elevation of multi-function reading center（Photographer, Trimont Image）

左上图：多功能阅览中心大楼梯入口连桥（陈畅 拍摄）Left upper: The grand stair entrance like a bridge of multi-function reading center（Photographer, Chen Chang）
右上图：多功能阅览中心局部（叁山影像 拍摄）Right upper: Multi-function reading center（Photographer, Trimont Image）
下图：多功能阅览中心"月亮门"入口空间（叁山影像 拍摄）Below: Moon gate entrance of multi-function reading center（Photographer, Trimont Image）
对页：多功能阅览中心东南立面局部夜景（陈畅 拍摄）Opposite: Southeast elevation of multi-function reading center at night（Photographer, Chen Chang）

下图:多功能阅览中心多功能厅(叁山影像 拍摄) Below: Multi-function hall of multi-function reading center (Photographer, Trimont Image)
对页:从多功能阅览中心看高等工程教育博物馆(陈畅 拍摄) Opposite: View of Higher Engineering Education Museum from multi-function reading center (Photographer, Chen Chang)

对页：高等工程教育博物馆主立面透视（陈畅 拍摄） Opposite: Main elevation of Higher Engineering Education Museum （Photographer, Chen Chang）

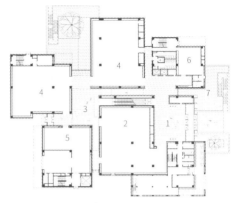

1	门厅	Lobby
2	通史展厅	General history exhibition hall
3	博物馆中庭	Courtyard of museum
4	临时展厅	Contemporary exhibition hall
5	库房	Storehouse
6	贵宾接待厅	VIP room
7	下沉庭院上空	Void above sunken garden

一层平面图 Ground floor plan

1	休息区	Lounge
2	展厅	Exhibition hall
3	博物馆中庭	Courtyard of museum
4	多功能厅	Multi-Function hall
5	室外平台	Outdoor platform

二层平面图 Second floor plan

1	休息区	Lounge
2	展厅	Exhibition hall
3	博物馆中庭	Courtyard of museum
4	展厅上空	Void above exhibition hall

三层平面图 Third floor plan

下图：局部透视（陈畅 拍摄）Below: Aspects (Photographer, Chen Chang)
对页：高等工程教育博物馆高低错落的体量关系（叁山影像 拍摄）Opposite: High and low volumes of Higher Engineering Education Museum (Photographer, Trimont Image)

1	藏品库房	Treasury storage	9	门厅	Lobby
2	中庭	Central Courtyard	10	下沉庭院	Sunken Garden
3	休息厅	Lounge	11	管理	Management Area
4	创意成果展厅	Exhibition Hall of Creative Achievements	12	走道	Walkway
5	科技成果展厅	Exhibition Hall of Scientific and Technological Achievements	13	过厅	Hallway
6	校史馆	University History Museum	14	专题展厅	Theme Exhibition Hall
7	通历展厅	General calendar Exhibition hall	15	临时展厅	Temporary Exhibition Hall
8	工程教育展厅	Engineering Education Exhibition Hall			

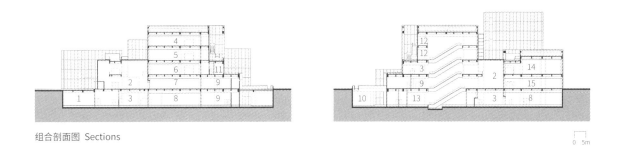
组合剖面图 Sections

下图：下沉庭院（叁山影像 拍摄）Below: Sunken garden（Photographer, Trimont Image）
对页：高等工程教育博物馆局部（陈畅 拍摄）Opposite:Higher Engineering Education Museum aspects（Photographer, Chen Chang）

龙岩市委党校
LONGYAN MUNICIPLE PARTY SCHOOL

项目地点 中国·福建·龙岩
项目规模 98347 m²
合 作 者 陈玲、黄斌毅、祝狄峰、朱祯毅、李聪、陈凤婷
项目时间 2013年设计，2020年竣工

Location Longyan, Fujian Province, China
Gross floor area 1058598.3 ft² (98347 m²)
Associates Chen Ling, Huang Binyi, Zhu Difeng, Zhu Zhenyi, Li Cong, Chen Fengting
Status Designed in 2013. Completed in 2020.

　　设计充分利用山水环境优势，把轴线的强调与因应地形的院落布局结合起来，创造一个灵动毓秀的校园空间，赋予整体环境浓烈的人文气质。屋面采用经过变换的坡屋面造型，错落有致，体现了闽西传统建筑韵味。方案通过对"场所、建筑、意境"三方面的创意设计，使得方案总体布局有机自然，空间形态多变，层次丰富，建筑风格雅致灵动，地上、地下综合利用，成为一座真正意义上的"绿色、现代、人文"党校。

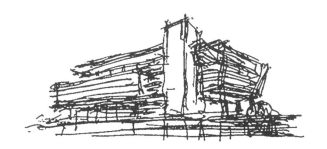

对页：中轴线鸟瞰（叁山影像 拍摄）
Opposite: Aerial view of main axis
(Photographer, Trimont Image)

对页：鸟瞰（陈畅 拍摄）Opposite: Aerial view （Photographer, Chen Chang）

The design makes full advantages of the landscape environment, combines the layout axis with the courtyard in response to the terrain, and creates a scenic campus space with a strong humanistic temperament. The transformed sloping roofs are randomly adopted and embodies the charm of traditional architecture in western Fujian. Through the creative design of "place, architecture and artistic conception", the general layout of the scheme is organic and natural, with varied spatial forms, differences in levels, elegant and smart architectural style, comprehensive utilization of levels both on ground and underground, and a real sense of "green, modern and humanistic" party school.

A1	大礼堂	Auditorium
A2	综合楼	Multiple-use building
A3	信息中心	Information center
A4, A5	教学楼	Teaching building
A6, A7, A8	阶梯教室	Lecture hall
B1	对外学员宿舍	Student dormitory (Outcome)
B2, B3, B4, B5	对内学员宿舍	Student dormitory (In school)
B6	食堂	Canteen

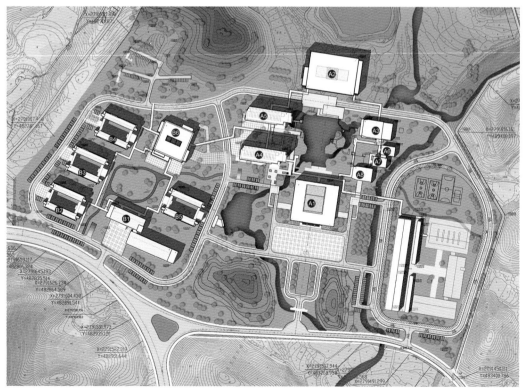

总平面图 General plan

下图：体育馆外景 （陈畅 拍摄） Below: Outside the indoor stadium （Photographer, Chen Chang）
对页：局部鸟瞰 （陈畅 拍摄） Opposite: Aerial view （Photographer, Chen Chang）

下图：礼堂室内（陈畅 拍摄）Below: Interior of auditorium （Photographer, Chen Chang）
对页：礼堂主入口外景（陈畅 拍摄）Opposite: Main entrance of auditorium （Photographer, Chen Chang）

下图：宿舍楼外景（陈畅 拍摄） Below: View of dormitory building （Photographer, Chen Chang）
对页：宿舍楼全景（陈畅 拍摄） Opposite: View of dormitory building （Photographer, Chen Chang）

上图：食堂一角（陈畅 拍摄）Top: One sight of canteen（Photographer, Chen Chang）
下图：教学楼架空层（陈畅 拍摄）Below: View of stilt floor in teaching building（Photographer, Chen Chang）
对页：宿舍楼外景（叁山影像 拍摄）Opposite: View of dormitory building（Photographer, Trimont Image）

青岛（红岛）铁路客站
QINGDAO (RED ISLAND) RAILWAY STATION

项目地点　中国·山东·青岛
项目规模　407380 m²
合作单位　中国铁路设计集团有限公司
合 作 者　于晨、金智洋、郭磊、严彦舟、陈立国、戚东炳、江畅、
　　　　　刘辛、孙铭、方炀、李嘉蓉、江钗、蒋美锋
项目时间　2016 年设计，2020 年竣工

Location　Qingdao, Shandong Province, China
Gross floor area　4385001.8 ft² (407380 m²)
Cooperators　CRDC
Associates　Yu Chen, Jin Zhiyang, Guo Lei, Yan Yanzhou, Chen Liguo,
　　　　　　　Qi Dongbing, Jiang Chang, Liu Xin, Sun Ming, Fang Yang,
　　　　　　　Li Jiarong, Jiang Chai, Jiang Meifeng
Status　Designed in 2016, Completed in 2020.

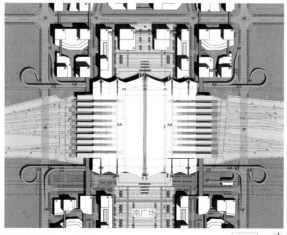

总平面图　General plan

对页：入口外景（陈畅 拍摄）
Opposite: Main entrance view
(Photographer, Chen Chang)

下图：设计草图——"梦由浪花，幻若飞檐" Below: Design sketch "Waves in dreams, flying into roofs"
对页：设计草图 Opposite: Design sketches

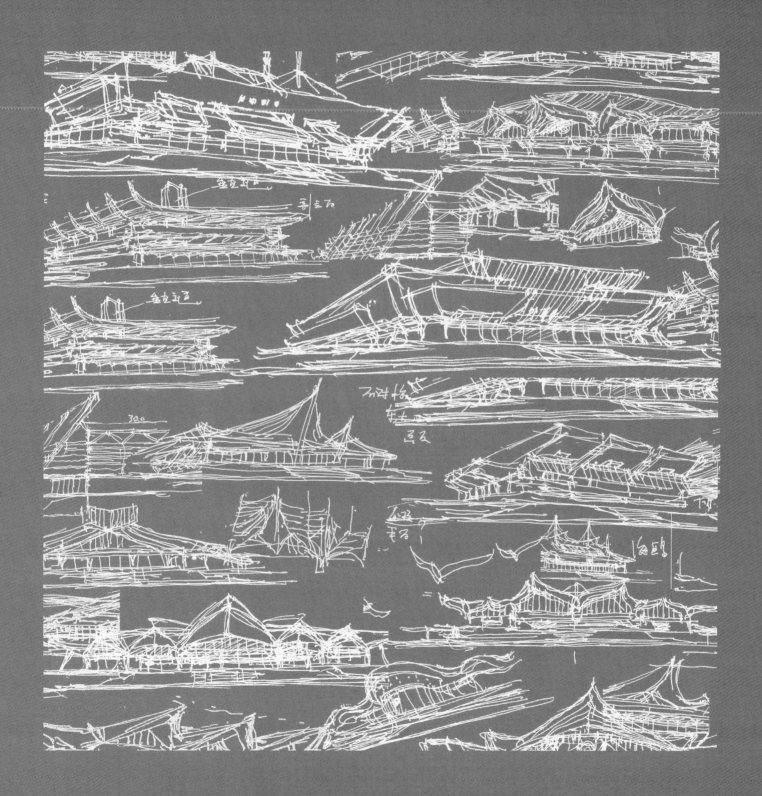

对页：全景鸟瞰（陈畅 拍摄）Opposite: Aerial panorama (Photographer, Chen Chang)

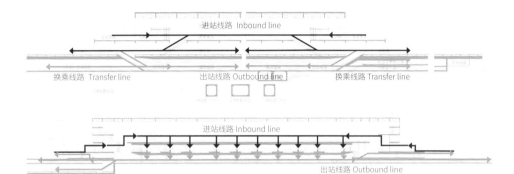

流线分析 Flow analysis

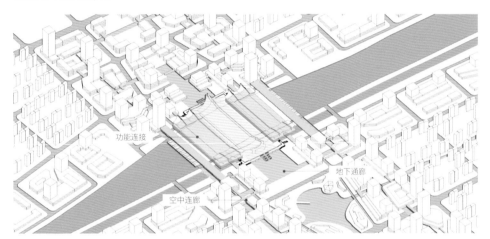

缝合城市 Stitching the city

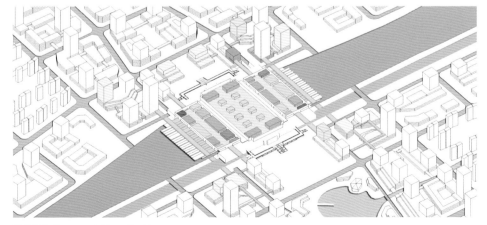

商业整合 Commercial investigation

青岛（红岛）铁路客站是济青高铁与青连铁路的始发站，未来将有三条线路通过，成为山东省东部的铁路枢纽。

The Qingdao (Red Island) Railway Station is the departure station of the Jinan-Qingdao High-speed Railway line (HSR) and of the Qingdao-Lianyungang Railway. In the future, there will be three traffic routes converging here, making this the eastern railway junction for Shandong Province.

对页：全景鸟瞰（陈畅 拍摄） Opposite: Aerial panorama (Photographer, Chen Chang)

1	地铁入口	Subway entrance
2	出租车上客区	Taxi access area
3	公交发车区	Bus departure area
4	社会车辆停放	Public vehicles parking area
5	集散广厅	Distribution hall
6	商业开发	Business development area

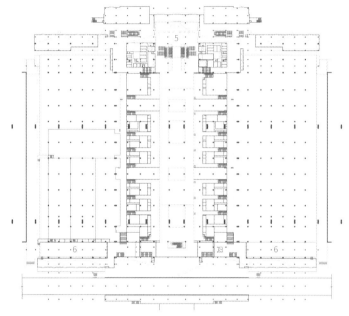

-7m 平面图 -7 m Level plan

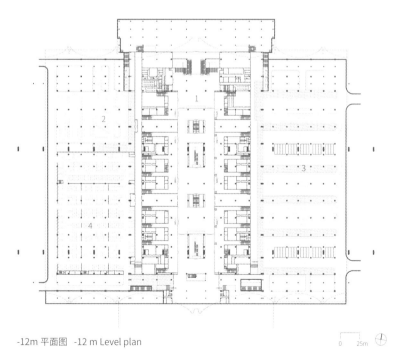

-12m 平面图 -12 m Level plan

对页：沿河外景（陈畅 拍摄）　Opposite: View along the river　(Photographer, Chen Chang)

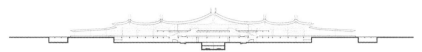

剖面图　Section

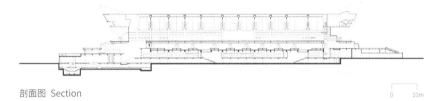

剖面图　Section

设计的愿景是：

第一，利用站房的地下通廊、高架步道的立体空间系统，将站房南北广场衔接，"缝合"因铁路而割裂的城市空间；

第二，将各个交通的换乘地点整合在一个交通枢纽里，实现"零换乘"；

第三，打造以高铁站房为主体的城市大型综合体；

第四，通过对周边建筑高度的控制以及空间轴线的塑造，让旅客饱览胶州湾美丽的海洋风光；

最后，以海浪为造型意象，表达滨海城市的气质特征。

The vision of this design is to make good use of the three-dimensional spaces, including the underground passage of the station buildings, the elevated pedestrian walkway, and the business center aboveground; to link the north and south squares and "sew together" the broken urban space disrupted by the railway.

Secondly, to gather the transfer points of various transportation systems into a single traffic junction to function as a "zero-transfer" interchange.

Thirdly, the project will involve the construction of a large-scale urban complex with the High-speed Railway station building serving as the main body.

Fourth, the design will provide travelers with full ocean views of beautiful Jiaozhou Bay through effective height controls of the peripheral buildings and a well-structured spatial axis.

Finally, using the shape and theme of a wave as the formative image to express the characteristic qualities of this coastal city.

下图及对页：建筑局部（陈畅 拍摄） Below and opposite: Aspects (Photographer, Chen Chang)

青岛（红岛）铁路客站 QINGDAO (RED ISLAND) RAILWAY STATION

下图及对页：落客车道及雨棚（陈畅 拍摄） Below and opposite: Drop off lane with shelter （Photographer, Chen Chang）

下图：进站大厅（陈畅 拍摄）Below: Station hall (Photographer, Chen Chang)
对页：候车厅（陈畅 拍摄）Opposite: Perspective of waiting hall (Photographer, Chen Chang)

下图：进站大厅（陈畅 拍摄）Below: Station hall（Photographer, Chen Chang）
对页：候车厅（陈畅 拍摄）Opposite: View of waiting hall（Photographer, Chen Chang）

青岛（红岛）铁路客站
QINGDAO (RED ISLAND) RAILWAY STATION

沿河外景（陈畅 拍摄）
View along the river (Photographer, Chen Chang)

苏州中学苏州湾校区
SUZHOU HIGH SCHOOL-SUZHOU BAY CAMPUS

项目地点　中国·江苏·苏州
项目规模　90307 m²
合作单位　中衡设计集团股份有限公司
建筑主创　程泰宁、刘翔华
合　作　者　殷建栋、吴妮娜、刘翔华、王岳峰、冯超、彭永雷、余新民、
　　　　　　许宁宁、阮珂、刘迟、何其乐、张道玉
项目时间　2016 年设计，2021 年竣工

Location Suzhou, Jiangsu Province, China
Gross floor area 972056.4 ft² (90307 m²)
Cooperators ARTS Group Co., Ltd.
Chief architect Cheng Taining, Liu Xianghua
Associates Yin Jiandong, Wu Nina, Liu Xianghua，Wang Yuefeng, Feng Chao, Peng Yonglei, Yu Xinmin, Xu Ningning, Ruan Ke, Liu Chi, He Qile, Zhang Daoyu
Status Designed in 2016, Completed in 2021.

　　项目位于吴江太湖新城，紧邻苏州市吴江区胜地湿地公园，场地现状为生态湿地，环境优美，水系交汇。

The project is located in Wujiang Taihu New Town, close to Suzhou Wujiang District Resort Wetland Park. The current site is ecological wetland and the environment is beautiful.

对页：南向鸟瞰（陈畅 拍摄）
Opposite: Aerial view from the south
（Photographer, Chen Chang）

对页：鸟瞰图（陈畅 拍摄）Opposite: Aerial view (Photographer, Chen Chang)

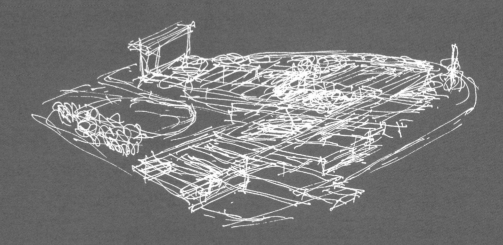

总平面图 General plan

整个校园动静分区明确，各功能建筑连贯整体的气势与交织的共享平台共同构成有机的整体。整体构架可分为"一网、四组团、多庭院"。

一网：共享活动平台自北向南如脉络状交织起整个校园的空间和流线。

四组团：分别为公共组团、教学组团、生活组团以及运动组团。

多庭院：建筑之间自然围合的院落，共享活动平台网状分布的庭院共同构成了校园内精彩纷呈的主题院落，营造了丰富的交流场所。

The whole campus is clearly divided into active and static zones. The coherent and integral momentum of each functional building and the interwoven sharing platform constitute an organic whole.The general layout can be introduced as "one net, four groups, multiple courtyards".

One net: The shared activity platform from north to south, like a net, interweave the space and circulations of the whole campus.

Four groups: public group, teaching group, life group and sports group.

Multiple courtyards: naturally enclosed courtyards between buildings, and courtyards distributed in the network of Shared activity platforms constitute the colorful theme courtyards in the campus, creating a rich place for communication.

对页：教学楼组团半鸟瞰（陈畅 拍摄）　Opposite: Aerial view of the teaching building cluster （Photographer, Chen Chang）

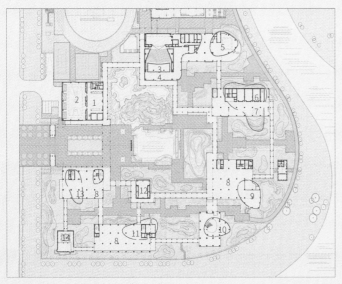

一层平面图 Ground floor plan

1	篮球场门厅	Lobby of basketball court
2	篮球场	Basketball court
3	报告厅	Lecture hall
4	报告厅前厅	Lobby of lecture hall
5	跆拳道	Aekwondo
6	教室	Classroom
7	室外小剧场	Outdoor studio theater
8	共享活动空间	Shared activity space
9	咖啡厅	Cafe
10	学习中心	Learning center
11	娱乐交流空间	Entertainment and communication space
12	180人报告厅	180ppl lecture hall
13	展示区	Display area
14	冥想花园	Meditation garden

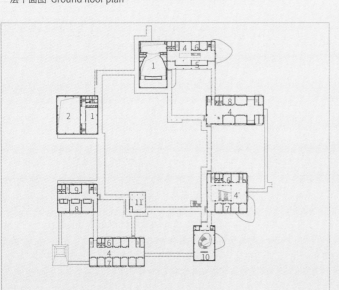

二层平面图 Second floor plan

1	舞蹈排练厅	Dance rehearsal hall
2	篮球场上空	Void above basketball court
3	报告厅上空	Void above lecture hall
4	活动区	Activity area
5	活动室	Activity room
6	机动教室	Flexible classroom
7	普通教室	Classroom
8	实验室	Lab
9	美术教室	Art classroom
10	学习中心	Learning center
11	梦工厂	Dream factory

下图：教学楼局部（陈畅 拍摄）　Below:Part of the teaching building（Photographer, Chen Chang）
对页：初中部教学楼（陈畅 拍摄）　Opposite: Teaching building of junior high school（Photographer, Chen Chang）

下图及对页：教学楼室内（陈畅 拍摄）　Below and opposite: Interior of teaching building （Photographer, Chen Chang）

杭州师范大学仓前校区
CANGQIAN CAMPUS OF HANGZHOU NORMAL UNIVERSITY

项目地点　中国·浙江·杭州
项目规模　407916 m²
合 作 者　王大鹏、柴敬、沈一凡、黄斌、王岳锋、周炎鑫、庄允锋、
　　　　　谢悦、陶韬、汤焱、祝容、胡晓明、贾秀颖
项目时间　2010 年设计，2018 年竣工

Location　Hangzhou, Zhejiang Province, China
Gross floor area　4390771 ft² (407916 m²)
Associates　Wang Dapeng, Chai Jing, Shen Yifan, Huang Bin,
　　　　　Wang Yuefeng, Zhou Yanxin, Zhuang Yunfeng, Xie Yue,
　　　　　Tao Tao, Tang Yan, Zhu Rong, Hu Xiaoming, Jia Xiuying
Status　Designed in 2010, Completed in 2018.

对页：**全景鸟瞰**（奥观建筑视觉 拍摄）
Opposite: **Aerial panorama**
(Photographer, AOGVISION)

对页：沿河鸟瞰（奥观建筑视觉 拍摄）Opposite: Aerial view along the river (Photographer, AOGVISION)

项目整体设计以"湿地书院"为创意定位，充分适应湿地肌理、传达传统书院的院落，其因地制宜、情景交融的规划布局，疏密有致的院落簇群与融传统与现代于一体的建筑单体，泛中心感的形态和适度复合的功能，创造了具有杭师大文化传承的开放式、生态化、园林式校园空间。建筑造型的构思主要来源于对中国传统文字构架和余杭历史建筑的认知，梁柱体系、坡屋顶、窗扇花格等原形要素的转换重现，形式在"似与不似之间"，既体现了灵动、雅致的杭州地域特点，又具大气简约的时代性。

Taking "Wetland Academy" as the creative positioning of design, the buildings are fully adapt to the wetland texture, representing the courtyards of traditional college in old times. The general layout adjusts measures to site conditions, the compound of the cluster that combine traditional and modern characteristics, the similarities in building form and its moderately composite function, have created the Hangzhou Normal University as an open, ecology, garden campus space. The conception of architectural modeling mainly comes from the cognition of the traditional Chinese character framework and historical building forms of Yuhang district, and the transformation and representation of the original elements such as beam and column system, sloping roof and sash-window lattice, with the form in-between "likeness and unlikeness", which not only reflects the elegant regional characteristics of Hangzhou, but also has the atmospheric characteristics of the time.

1	新能源学院	School of New Energy
2	材化学院	School of Materials and Chemical Engineering
3	公共实验楼一	Public Laboratory Building No.1
4	公共实验楼二	Public Laboratory Building No.2
5	数理类教学楼	Mathematics and Physics Teaching Building
6	军工类实验楼	Military Industry Laboratory Building
7	遥感技术学院	School of Remote Sensing Technology
8	国际服务外包学院	School of International Service Engineering
9	计算机学院	School of Computer Science
10	公共教学楼二	Public Teaching Building No.2
11	图书馆分馆	Library Branch
12	公共教学楼一	Public Teaching Building No.1
13	工科实验楼一	Engineering laboratory Building No.1
14	工科实验楼二	Engineering laboratory Building No.2
15	专家公寓	Expert's Apartment
16	教师公寓一	Teacher's Apartment No.1
17	教师公寓二	Teacher's No.2
18	研究生公寓	Graduate Student Apartment
19	本科生公寓二	Undergraduate Student Apartment No.2
20	本科生公寓一	Undergraduate Student Apartment No.1
21	学生活动中心	Student Center
22	物业管理用房	Property Management Room
23	食堂	Canteen
24	产学研综合体	Industry-University-Research Complex

总平面图　General plan

下图及对页：公共教学楼外景（奥观建筑视觉 拍摄） Below and opposite: Public teaching building (Photographer, AOGVISION)

下图：公共实验楼局部外景（奥观建筑视觉 拍摄）
Below: View of public laboratory building （Photographer, AOGVISION）
对页：公共教学楼外景（奥观建筑视觉 拍摄）
Opposite: View of public teaching building （Photographer, AOGVISION）

下图：国际服务学院外景（奥观建筑视觉 拍摄）Below: View of international service college (Photographer, AOGVISION)
对页：图书馆外景（奥观建筑视觉 拍摄）Opposite: View of library (Photographer, AOGVISION)

下图：学生活动中心外景（奥观建筑视觉 拍摄）Below: View of student center（Photographer, AOGVISION）
对页：工科实验楼外景（奥观建筑视觉 拍摄）Opposite: View of engineering laboratory building（Photographer, AOGVISION）

下图：食堂屋顶露台空间（奥观建筑视觉 拍摄）Below: Canteen roof terrace space（Photographer, AOGVISION）
对页：食堂（陈畅 拍摄）Opposite: Canteen（Photographer, Chenchang）

下图：架空空间（奥观建筑视觉 拍摄）　Below: Elevated space（Photographer, AOGVISION）
对页：连廊空间（奥观建筑视觉 拍摄）　Opposite: Corridor space（Photographer, AOGVISION）

杭州西站及城市综合体
HANGZHOU WEST RAILWAY STATION COMPLEX

项目地点　中国·浙江·杭州
项目规模　510000 m²
合作单位　中铁第四勘察设计院集团有限公司
　　　　　杭州城市规划设计研究院
合 作 者　上海公司：于晨、戚东炳、金智洋、江畅、方炀
　　　　　杭州公司：殷建栋、严彦舟、陈立国、张昊楠、邱培昕
　　　　　云门深化设计：薄宏涛、郑庆丰
　　　　　铁四院方案设计：盛晖、李立、殷炜
项目时间　2017 年设计，国际招标中标，施工中

Location　Hangzhou, Zhejiang Province, China
Gross floor area　5489594.3 ft² (510000 m²)
Cooperators　China Railway Siyuan Survey and Design Group Co.,
　　　　　LTD(CRFSDI)
　　　　　Hangzhou Urban Planning and Design Institute
Associates　CCTN Shanghai: Yu Chen, Qi dongbing, Jin Zhiyang,
　　　　　Jiang Chang, Fang Yang
　　　　　CCTN Hanghzou: Yin Jian dong, Yan Yanzhou,
　　　　　Chen Liguo, Zhang Haonan, Qiu Peixin
　　　　　Cloud gate design: Bo Hongtao, Zheng Qingfeng
　　　　　CRFSDI: Sheng Hui, Li Li, Yin Wei
Status　Designed in 2017, won bid in international competition,
　　　　　under construction.

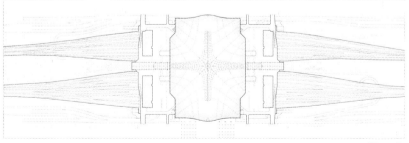

总平面图　General plan

对页：云门夜景
Opposite: Cloud gate perspective at night

鸟瞰图 Aerial view

整体设计以打造"站城融合发展"、城市未来生活的典范区为目标。结合整体"云之城"的理念,我们将站房综合体设计融入整体环境考虑,站房与城市综合体多层次、多维度的建立联系,将站融于城,将城合于站,打造站城高度融合的超级 TOD。建立火车站与余杭塘河,以及寡山、吴山,"看"与"被看"的相互关系,形成"呼应山水,沟通南北"的自然生态格局。

The general plan aims to create a "Development of station-city synergy", a model project for future urban life. In combination with the concept of "cloud city", we integrate the design of station complex into the city environment consideration, establish multi-level and multi-dimensional connection between station and urban complex, stitch the station into the city, and the city into the station vice versa, and create a super TOD with highly integrated station and city. The interrelationship

东入口全景 East entrance perspective

下图：杭州西 - 爆炸图 - 交通流线 Below: Traffic circulations
对页：剖透视——各类交通的零换乘 Opposite: Sectional perspective—"zero" transfer among all means of transportation

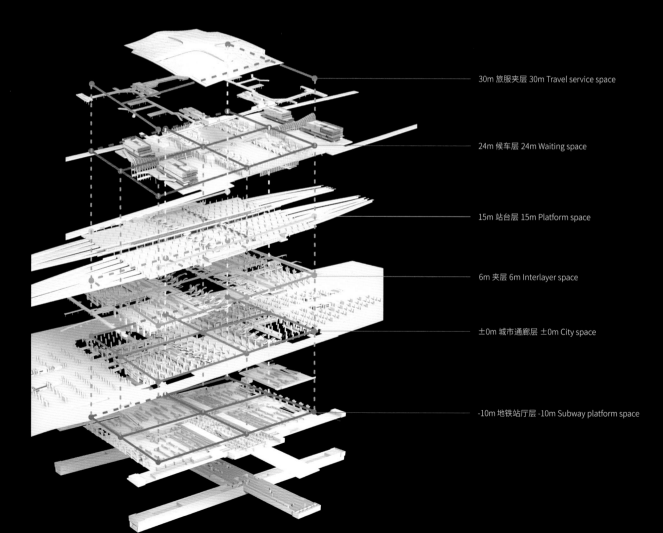

30m 旅服夹层 30m Travel service space
24m 候车层 24m Waiting space
15m 站台层 15m Platform space
6m 夹层 6m Interlayer space
±0m 城市通廊层 ±0m City space
-10m 地铁站厅层 -10m Subway platform space

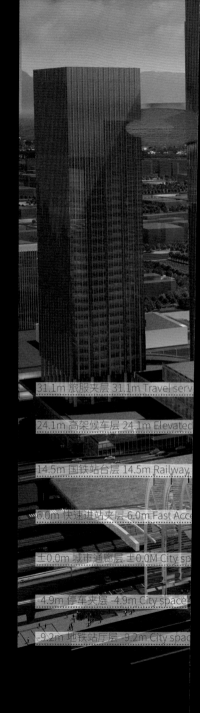

31.1m 旅服夹层 31.1m Travel service
24.1m 高架候车层 24.1m Elevated
14.5m 国铁站台层 14.5m Railway
6.0m 快速进站夹层 6.0m Fast Access
±0.0m 城市通廊层 ±0.0M City space
-4.9m 停车夹层 -4.9m City space
-9.2m 地铁站厅层 -9.2m City space

下图：设计草图 Below: Design sketch
对页：云门正透视 Opposite: Cloud gate perspective

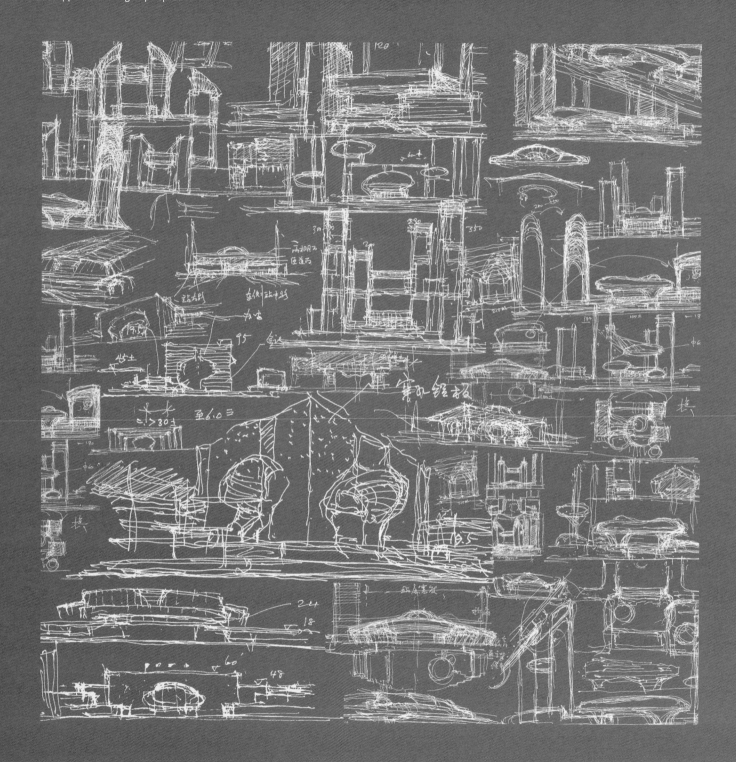

对页：云门——城市客厅 Opposite: Cloud gate—the city-hall in station

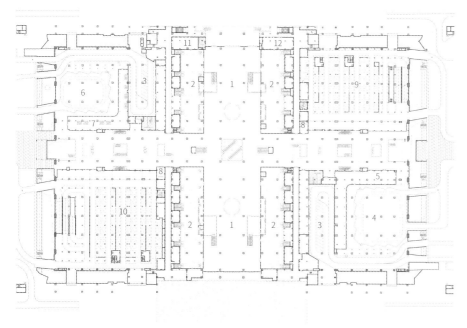

1	城市通廊	City corridor
2	出站厅	Outbound hall
3	出租车场	Taxi parking
4	公交车场	City bus parking
5	公交车候车区	City bus waiting area
6	大巴车场	Shuttle bus parking
7	大巴车候车区	Shuttle bus waiting area
8	售票区	Tickets
9	停车库（地方）	Parking area for city vehicles
10	停车库（国铁）	Parking area for national railways
11	预留办公区	Reserved offices
12	消防、综合控制区	Fire control/integrated control area

地面广场层（0.000 标高）平面 Ground floor plaza plan (±0.000m)

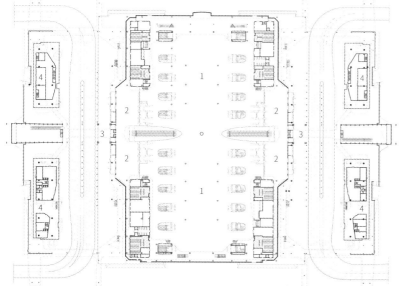

1	候车厅	Waiting hall
2	进站厅	Inbound hall
3	落客平台	Passengers' platform
4	上盖开发	Railway hub development

高架候车层（24.100 标高）平面 Elevated waiting floor plan (+24.100m)

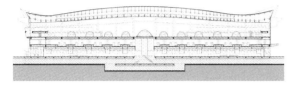

剖面图 Section

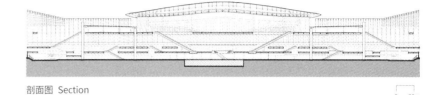

剖面图 Section

0　20m

上图：云门落客区平台 Top: Passenger's platform at the cloud gate
下图：综合开发落客区 Below: Multifunctional guest area
对页：落客平台透视 Opposite: Passenger's platform perspective

下图：云门共享空间 Below: Shared space at the cloud gate
对图：云门入口内视 Opposite: Entrance of the cloud gate

　　站房部分将打造成"云端站房"，提出"云谷""云路""云厅"的核心理念。通过"云谷"（站场拉开空间）解决站与城复杂的交通组织，通过"云路"打造中央快速进站系统，通过"云厅"将站与城的功能更好地融合起来。

The station house part is made as a "cloud station house", and follows the core concepts of "cloud valley", "cloud road" and "cloud hall".

Through "cloud valley" (an open station yard space), to solve the complex traffic organization between the station and the city; Through "cloud road", to build the central quick transport system; Through "cloud hall", the functions of the station and the city will be better integrated.

下图：云谷大厅透视 Below: Cloud valley hall perspective
对页：候车大厅透视 Opposite: Waiting hall perspective

厦门一场两馆、新会展中心城市设计及建筑概念设计方案
CONCEPTUAL URBAN DESIGN PROPOSAL FOR THE OUTDOOR STADIUM, INDOOR GYMNASIUM AND NATATORIUM AND NEW CONVENTION CENTER IN XIAMEN

项目地点　中国·福建·厦门
项目规模　城市设计 636.62 万 m^2
　　　　　　新会展中心 106.4 万 m^2
　　　　　　新体育中心 42 万 m^2
合作单位　东南大学建筑设计研究院有限公司
　　　　　　西班牙波菲建筑设计有限责任公司
合 作 者　东南大学建筑设计研究院有限公司团队：韩冬青、高庆辉、王志刚等
　　　　　　筑境设计团队：周旭宏、殷建栋、黄斌毅、张莹、刘翔华等
　　　　　　波菲建筑设计有限责任公司团队：Ricardo Bofill
项目时间　2019 年设计，国际招标中标，作为总控单位设计深化。施工中

Location　Xiamen, Fujian Province, China
Gross floor area　Urban Design 6366200 m^2
　　　　　　New Convention and Exhibition Center 1064000 m^2
　　　　　　New Sports Center 420000 m^2
Cooperators　Architects & Engineers Co.,LTD. of Southeast University
　　　　　　RBTA Co.,LTD.
Associates　Architects & Engineers Co.,LTD. of Southeast University Team: Han Dongqing, Gao Qinghui, Wang Zhigang, etc.
　　　　　　CCTN Team: Zhou Xuhong, Yin Jiandong, Huang Binyi, Zhang Ying, Liu Xianghua etc.
　　　　　　RBTA Co.,LTD. Team: Ricardo Bofill
Status　Won bid as executive managing firm in 2019. Under construction.

对页：体育会展新城鸟瞰
Opposite: Aerial view of the new city of sports and exhibition

下页：体育会展新城鸟瞰 Next page: Aerial view of the new city of sports and exhibition

厦门市"新会展中心"项目位于厦门市翔安区滨海东大道以南、东界路以东、扬帆公园以西、滨海花园大道以北地块。主要建设内容包括展览中心，会议中心，滨海商业及配套、地下停车场等。总建筑面积约106万 m²。

结合厦门当地气候特色，新会展中心提出鱼骨+庭院的布局模式，会展中央设计开放庭院，中心节点设计圆形立体景观，并沿庭院植入餐饮、洽谈功能，打造流线清晰、环境优美的创新型会展建筑。会展中心造型通过对大厝文化中"燕尾、曲脊、重檐"以及海浪、白鹭等元素的抽象提炼，以富有层次的屋顶、典雅的建筑细节营造会展中心"又见大厝现蓝海，恍若白鹭落九天"的诗意氛围。

Xiamen "New Convention Center" project is located in the south of Binhai East Avenue, East of Dongjie Road, west of Yangfan Park and north of Binhai Garden Avenue. The main construction includes exhibition center, convention center, waterfront business and supporting facilities, underground parking lot, etc. The total construction area is about 1.06 million square meters.

Considering the local climate characteristics of Xiamen, the general layout of new convention center proposed the 'fishbone + courtyard' mode. The open courtyard is designed in the center of the exhibition center, and the circular and three-dimensional landscape is designed at the center node. Besides, catering and other functions are embedded in the courtyard to create an innovative exhibition building with clear circulation and beautiful environment. By extracting elements from the Dacuo culture, forms such as dovetail, ridge and double eave roofs and images such as waves and egrets are adopted to the building form. Elegant architectural details create a convention center against the wind and waves, creating a magnificent scene along the seaside, to achieve the poetic atmosphere of "seeing the big building standing inside blue sea, like egrets fall from the sky".

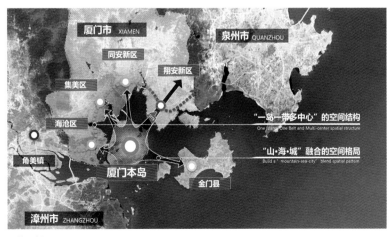 

区位分析 Location analysis

区块总指标		
地块总开发量	636.62万㎡	
其中	住宅面积	208.32万㎡
区块编号	区块名称	开发量（万㎡）	容积率
A	新体育中心片区	42.0	0.6
B	新会展中心片区	85.5	1.0
C	中轴滨海区	24.8	2.6
D	中轴南区	122.1	6.8
E	中轴北区	112.8	4.9
F	滨海东大道片区	77.9	4.0
G	居住区	171.6	3.5

总平面图 General plan

厦门一场两馆、新会展中心城市设计及建筑概念设计方案
CONCEPTUAL URBAN DESIGN PROPOSAL FOR THE OUTDOOR STADIUM, INDOOR GYMNASIUM AND NATATORIUM AND NEW CONVENTION CENTER IN XIAMEN

下图：体育会展新城滨海天际线 Below: Marina skyline of the new city of sports and exhibition

厦门新会展中心
THE NEW CONVENTION CENTER OF XIAMEN

项目地点　中国·福建·厦门
项目规模　展览中心总建筑面积 83.29 万 m²
　　　　　会议中心总建筑面积 23.11 万 m²
合 作 者　杭州：殷建栋、张莹、刘翔华、彭永雷、高凌琪、谢晓峰、
　　　　　古振强、邱培昕、赵晨曦、高良杰
　　　　　上海：周旭宏、杨嘉、葛金玉、毛磊、郑从涛、李雪晗、
　　　　　吴柳培、颜伊岚
项目时间　2019 年中标，施工中

Location　Xiamen, Fujian Province, China
Gross floor area　New Exhibition Center 832900 m²
　　　　　　　　　　New Convention Center 231100 m²
Associates　CCTN Hangzhou: Yin Jiandong, Zhang Ying, Liu Xianghua,
　　　　　　　Peng Yonglei, Gao Lingqi, Xie Xiaofeng, Gu Zhenqiang,
　　　　　　　Qiu Peixin, Zhao Chenxi, Gao Liangjie
　　　　　　　CCTN Shanghai: Zhou Xuhong, Yang Jia, Ge Jinyu, Mao lei,
　　　　　　　Zheng Congtao, Li Xuehan, Wu Liupei, Yan Yilan
Status　Won bidding in 2019, under construction.

对页：设计草图　Opposite: Design sketch

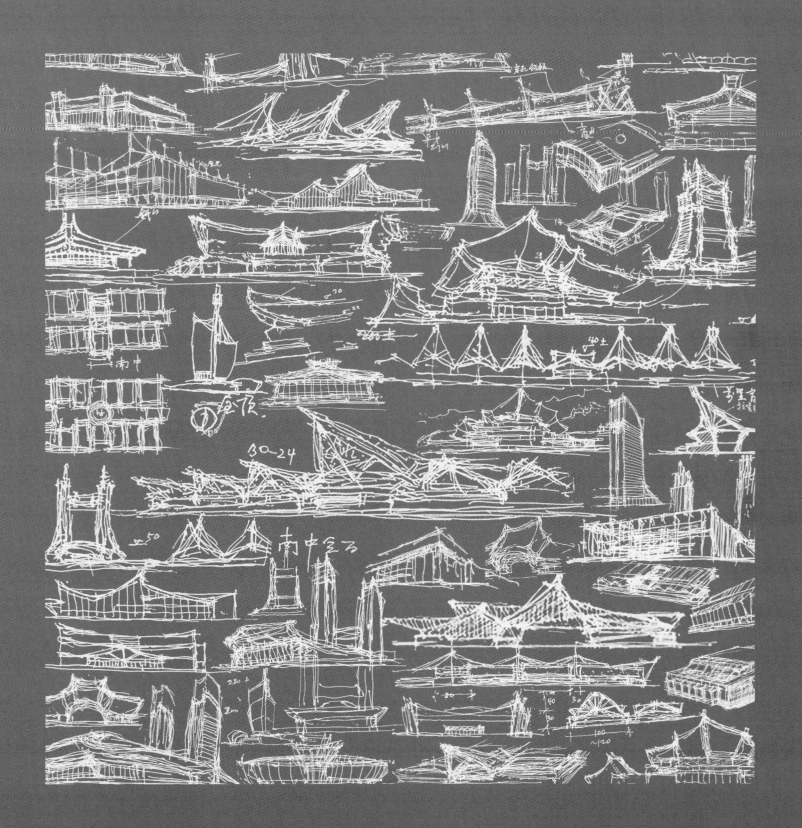

对页：会展中心整体鸟瞰 Opposite: Aerial view of the conference center

厦门新会展中心位于厦门东部体育会展新城的沿海核心地带，凭海临风，坐望本岛，是集展览、会议、酒店、休闲等多功能为一体的复合型会展综合体。旨在以区域会展产业经济为引擎，以助力打造"活力海滨"为目标，开启区域会展产业发展新纪元，擎动厦门跨岛经济实现新突破、新发展。

The Xiamen New Convention Center is located in the coastal core of the eastern sports and exhibition new town of Xiamen. It is a complex integrating exhibition, conference, hotel, leisure and other functions, with the seaside view and the blowing sea wind. The aim is to take the regional exhibition industry as the engine, to help build a "vibrant seaside zone", to start a new era of regional exhibition industry development, and to drive Xiamen to achieve new economical breakthroughs and developments crossing islands.

对页：会展中心滨海透视 Opposite: Seaside Perspective of the conference center

位于一线滨海的厦门会议中心是厦门新会展中心实现"会中带展、展中带会"集聚化发展的核心组成部分，是厦门市以做大做强本市会展业为导向的对标国际的大型会议中心。

厦门会议中心以闽南大厝、九天白鹭为其设计意向，造型设计汲取闽南大厝之精髓，传承燕尾曲脊之神韵，用现代建筑手法诠释厦门地域文化特色——利用端庄大气的三段式古典建筑构成手法，灵动有力、张弛有度的双曲金属屋面，轻盈剔透、细致考究的建筑幕墙，去诠释这个面朝大海，属于厦门、更属于世界的当代大型会议中心。

The Conference Center, located along the coastal line, is the core component of Xiamen New Convention Center to help the building realize an agglomeration development of "exhibition within conference, conference within exhibition". It also follows the orientation of Xiamen to enlarge and strengthen the convention and exhibition industry of the city.

Taking the roof form of houses in southern Fujian area and egret wings as image, the design absorbs the essence of house form in southern Fujian, inherits the verve of the dovail and the ridge, and interprets the regional cultural characteristics of Xiamen with modern architectural techniques, such as to use dignified three-section classical architectural composition techniques, to use the hyperbolic metal roof with dynamic strength and relaxation; and to use the light and delicate curtain wall-- all of which suggest that the convention center facing the sea is an international center of the world.

对页：展览中心鸟瞰图 Opposite: Aerial view of the exhibition center

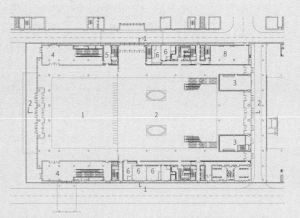

1	检录处	Login area
2	登录厅	Arriving hall
3	商务中心	Business center
4	公共厅	Public hall
5	电梯厅	Elevator lobby
6	服务用房	Service rooms
7	卫生间	Washroom
8	设备用房	Equipment rooms
9	贵宾室	VIP room

登录厅一层平面图 Ground floor plan of Entrance hall

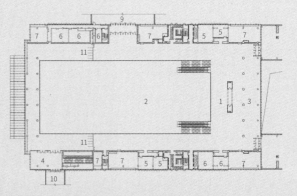

1	服务台	Service desk
2	登录厅上空	Void above arriving hall
3	休闲观景区域	Leisure and viewing area
4	公共厅	Public hall
5	展览配套	Facilities of exhibition
6	服务用房	Service rooms
7	设备用房	Equipment rooms
8	卫生间	Washroom
9	接展厅连廊	Corridor between exhibition halls
10	接会议中心连廊	Corridor link to conference center
11	检录处	Login

登录厅二层平面图 Second floor plan of Entrance hall

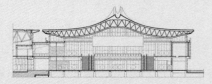

剖面图 Section

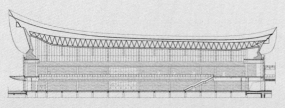

剖面图 Section

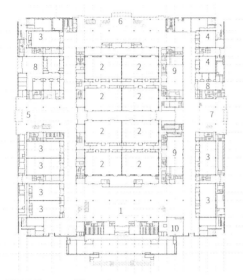

一层平面图 Ground floor plan

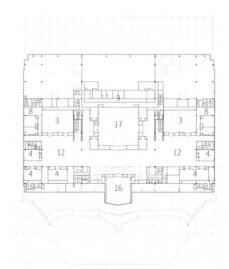

二层平面图 Second floor plan

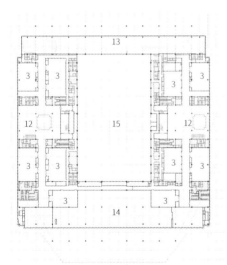

三层平面图 Third floor plan

四层平面图 Fourth floor plan

1	南入口大厅	South entrance hall	7	东入口大厅	East entrance hall	
2	大型会议室	Convention room	8	贵宾休息区	Vip lounge	
3	中型会议室	Medium conference room	9	厨房	Kitchen	
4	小型会议室	Creating room	10	商务中心	Business center	
5	西入口大厅	West entrance hall	11	超大型多功能会议室	Multi-function conference room	
6	北入口大厅	North entrance hall	12	茶歇区	Tea break area	
13	北入口大厅上空	Void above north entrance hall				
14	南入口大厅上空	Void above south entrance hall				
15	超大型多功能会议室上空	Void above multi-function conference room				
16	观海会议厅	Sea view meeting room				
17	国际会议厅	International meeting room				

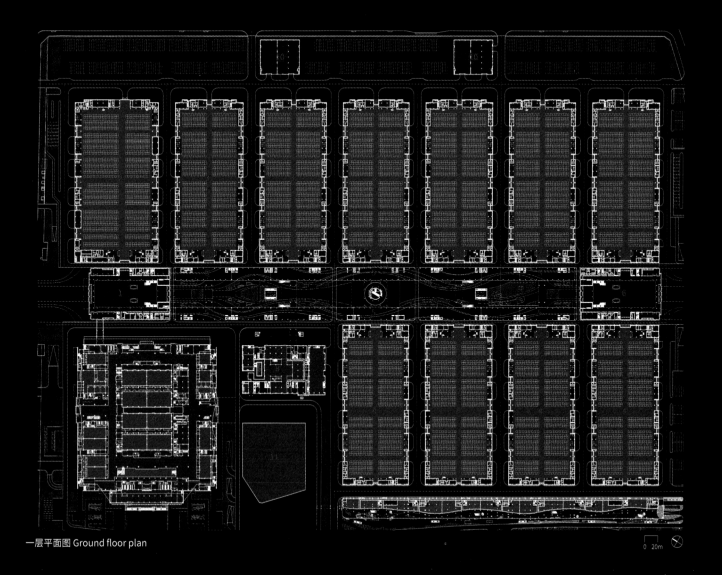

1	西登录厅	West entrance hall	7	滨海配套	Facilities along the seaside
2	3万展厅	30000 ㎡ Exhibition hall	8	能源中心	Resources center
3	2.4万北展厅	24000 ㎡ Exhibition hall	9	会议中心	Conference center
4	2.4万南展厅	24000 ㎡ Exhibition hall	10	仓库	Garage
5	中央廊道	Central corridor	11	酒店	Hotel
6	东登录厅	East entrance hall			

一层平面图 Ground floor plan

对页：会议中心滨海透视 Opposite: Conference center waterfront perspective

1　国际会议厅　　　International Conference Hall
2　观海会议厅　　　Sea-view Conference Hall
3　超大型多功能厅　Super Multi-function Hall
4　北入口大厅　　　North Entrance Lobby
5　大型会议室　　　Large Conference Room
6　南入口大厅　　　South Entrance Lobby
7　西入口大厅　　　West Entrance Lobby
8　东入口大厅　　　East Entrance Lobby

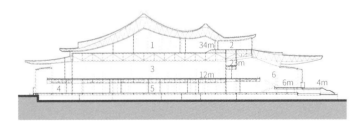

剖面图 Section

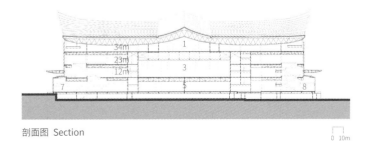

剖面图 Section

下图：会议中心滨海透视 Below: Conference center waterfront perspective
对页：会议中心透视图 Opposite: Conference center perspective

会议中心景观设计旨在与建筑形成统一协调的对外形象，通过景观设计烘托会议中心的独特氛围，同时也为市民提供可共享的城市空间。

The landscape design of the conference center aims to form a unified and coordinated image of the whole, to highlight the unique atmosphere of the conference center through landscape design, and to provide urban space for citizens to share.

线、曲面元素的连续、镜像、叠加、肌理等设计手法使空间轻盈灵动，充满韵律与节奏感。同时室内设计还提炼闽南地域文化特征中的大厝、燕尾脊，手工艺代表中的漆线雕、影雕、德化瓷、莆田木雕等作为肌理及纹样运用到空间中，以彰显地域文化特色。

The interior design continues the design concept of exterior form, transforming the abstractions of house form in southern Fujian and egrets into design language and applying them to the space. Through the continuous, mirror, superposition and texture design techniques of curve and surface elements, the space is light and lively, full of rhythm. At the same time, the interior design also refines the house and dovetail ridge in the regional cultural characteristics of southern Fujian, and the lacquer line carving, shadow carving, Dehua porcelain and Putian wood carving in the handicraft representative are applied into the space as texture and pattern, so as to highlight the regional cultural characteristics.

北京新国展二、三期城市设计及建筑概念设计方案
INTERNATIONAL COMPETITION PROPOSAL OF URBAN DESIGN AND PHASE II ARCHITECTURE DESIGN OF NEW BEIJING INTERNATIONAL EXHIBITION CENTER

项目地点　中国·北京
项目规模　813575m²
合作单位　东南大学建筑设计研究院有限公司
　　　　　南京东南大学城市规划设计院有限公司
合 作 者　东南大学建筑设计研究院有限公司：韩冬青、王志刚
　　　　　筑境设计团队：周旭宏、殷建栋、刘翔华、邱培昕
　　　　　东南大学城市规划设计研究院有限公司：朱彦东
项目时间　2019 年设计，国际招标入围

Location　Beijing, China
Gross floor area　8757248 ft² (813575m²)
Cooperators　Architects & Engineers Co.,LTD. of Southeast University
　　　　　Urban Planning & Design Institute of S.E.U
Associates　Architects & Engineers Co.,LTD. of Southeast University
　　　　　Team：Han Dongqin, Wang Zhigang
　　　　　CCTN Team：Zhou Xuhong, Yin Jiandong, Liu Xianghua, Qiu Peixin
　　　　　Urban Planning & Design Institute of S.E.U Team：Zhu Yandong
Status　Designed in 2019, enter the finalist of international bidding.

对页：西南侧鸟瞰
Opposite: Aerial view from the southwest

对页：正东鸟瞰 Opposite: Aerial view from the east

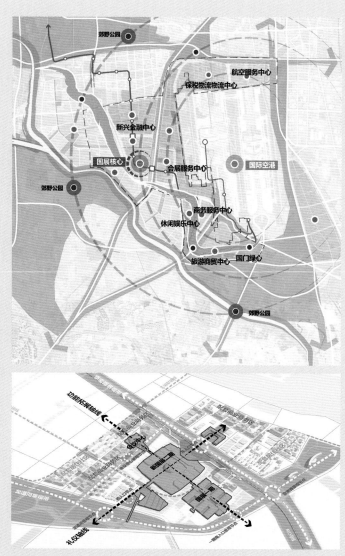

区位分析 Location analysis

中国庭院，创新会展

新国展采用有别于传统线性展览模式的庭院+向心式布局，各展厅环绕庭院成向心式布置。营造具有"中国院子"特色、打造更具有活力和展览氛围的花园会展。

交通环廊串接各个展厅，相对传统的线性展览模式，更能突出室内外空间的复合效应，打造更为丰富生动的展览空间。

未来展厅空间开敞、奇幻，可融合前沿科技，赋予观众以身临其境的观展体验。多功能主题展厅可展览、会议兼用。

闭展期间，未来展厅、多功能主题展厅以及环廊商业都可对外开放。形成足够体量和高度复合的业态，完全能够支撑展后的独立使用，打造不谢幕的城市公共活动空间。

Chinese Courtyard, Innovative Exhibition

Different from the traditional linear exhibition mode, the new BIEC adopts a courtyard + centripetal layout, with each exhibition hall surrounding the courtyard, thus to create a garden exhibition with the characteristics of "Chinese courtyard" and a more dynamic atmosphere.

The traffic ring corridor connects each exhibition hall, which can highlight the compound effect of indoor and outdoor space and create a more rich and vivid exhibition space compared with the traditional linear exhibition mode.

The exhibition space of the future is open and fantastical, which can integrate cutting-edge technology and give the audience an immersive exhibition experience. Multifunctional theme exhibition hall, exhibition and conference.

During the closing period, the future exhibition hall, the multi-purpose theme exhibition hall and the ring corridor business can be opened to the public. Form enough volume and highly complex format, fully able to support the independent use after the exhibition, to create the urban public activity space.

对页：东南鸟瞰　Opposite: Aerial view from the southeast

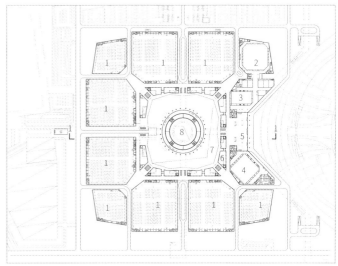

1	展厅	Exhibition hall
2	会议兼展厅	Meeting room with exhibition
3	大宴会厅	Grand banquet hall
4	大会场	Grand meeting room
5	安检大厅	Safety check
6	商业	Commercial space
7	庭院上空	Void above courtyard
8	常设展厅上空	Void above exhibition hall

一层平面图　Ground floor plan

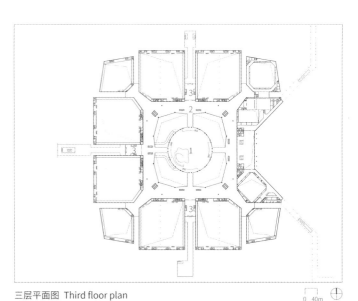

1	展厅	Exhibition hall
2	走廊	Corridor
3	登录厅	Entrance hall

三层平面图　Third floor plan

0　40m

230

下图：正入口立面透视　　Below：Entrance perspective

花开盛世，国色牡丹

新国展建筑造型通过对国花牡丹外形的抽象提炼，表达了"花开盛世、国色牡丹"的建筑意象。达到功能与空间、建构与意象的完美统一，具有强烈的视觉冲击力。

Blossoming Flower, a National Peony

The architectural form of the new BIEC expresses the architectural image of "blossoming flower and national peony" by extracting and applying the abstract shape of the national flower peony, to achieve the perfect unity of function and space, construction and image, with a strong visual impact.

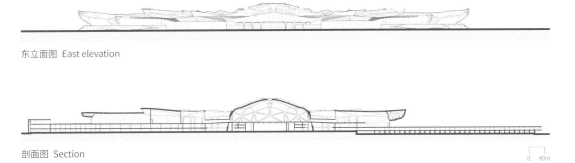

东立面图 East elevation

剖面图 Section

下图：近景透视　Below: Close view
对页：正入口近景透视　Opposite: Entrance perspective

下图：未来展厅入口 Below: Entrance of the future exhibition hall
对页：主登录厅 Opposite: Main lobby

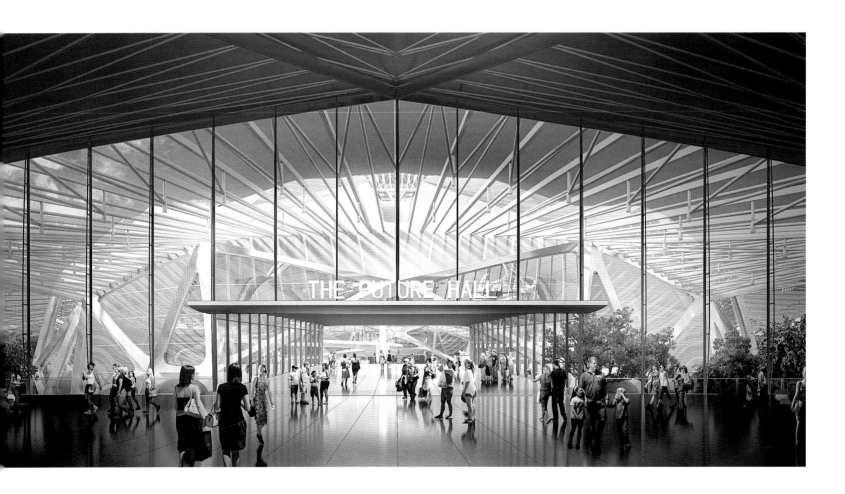

下图：标准展厅 Below: Exhibition hall
对页：交通环廊 Opposite: Ring corridor

下图：未来展厅 Below: Future exhibition hall
对页：中央露天庭院 Opposite: Center outdoor courtyard

中央庭院引入适当的室外科技展览，庭院也成为展厅。科技体验、艺术中心、展览博物馆等建筑功能向户外延伸，以吸引更多的人流，可打造不谢幕的城市公共活动空间。

The central courtyard also serves as an exhibition hall for outdoor science and technology exhibitions. Science and technology experience, art center, exhibition museum and other functions are extended to the outdoors, which can attract more people and create an 24-hour urban public activity space.

中央庭院为商务客户、科创客群、家庭亲子、年轻聚会等公众使用者，提供开放、包容、充满阳光、自然、具有活力的地标式中央绿地。

The central courtyard provides an open, inclusive, sunny, natural and dynamic landmark central green space for business customers, science and technology groups, families, young parties and other public users.

中央庭院植入绿荫光庭休闲主题咖啡区、商务洽谈咖啡茶座、室外花园餐厅、室外书吧、室外艺廊等，营造尺度宜人的开放式庭院。

In the central courtyard, there are leisure theme coffee area with green shade and light courtyard, coffee and tea house for business negotiation, outdoor garden restaurant, outdoor book bar, outdoor art gallery and other areas, creating an open courtyard with pleasant scale.

国展采用庭院+向心式布局模式。中央庭院以绽放花园为景观概念，延续建筑花开盛世的空间形态，形成整体流动而舒展开放的自然绿色中庭，打造环境优美的创新型花园会展。

The exhibition center adopts a layout mode as courtyard + centripetal. The central courtyard takes the blooming garden as the landscape concept, continues the space form of the blooming age of the whole building, forms a flowing, open and natural green atrium, and creates an innovative garden exhibition with beautiful environment.

东北夜景鸟瞰
Aerial view from the northeast at night

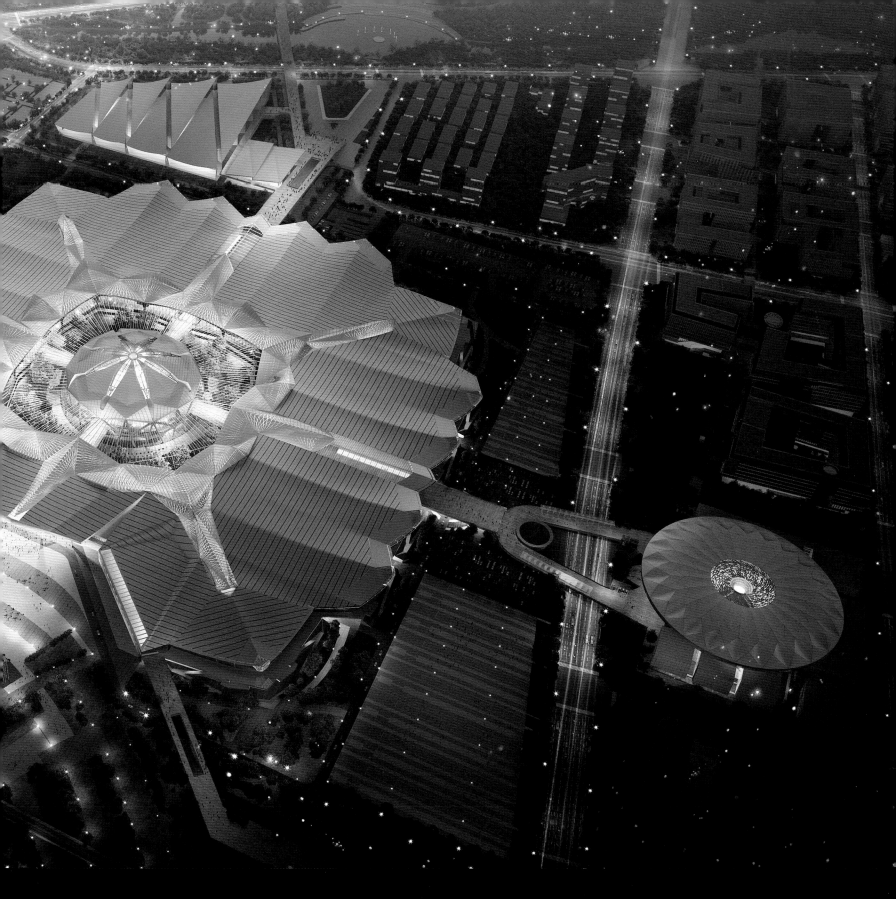

国深博物馆建筑设计方案
GUOSHEN MUSEUM PROPOSAL

项目地点　中国·广东·深圳
项目规模　127898 m²
合 作 者　王大鹏、汪毅、张潇羽、马聃、廖慧雯、施妙佳、赵效鹏、
　　　　　　缪斯、杜啸、李文亚、方曼雪
项目时间　2019 年设计，国际竞标方案

Location Shenzhen, Guangdong Province, China
Gross floor area 1376682.6 ft² (127898 m²)
Associates Wang Dapeng, Wang Yi, Zhang Xiaoyu, Ma Dan,
　　　　　　Liao Huiwen, Shi Miaojia, Zhao Xiaopeng, Miao Si,
　　　　　　Du Xiao, Li Wenya, Fang Manxue
Status Designed in 2019, not implemented.

总平面图　General plan

对页：沿前湾公园
Opposite: Perspective along the Qianwan Park

对页：设计草图 Opposite: Design sketch

国家博物馆的定位与前海海湾浪漫开放的气质相融合，形成多义独特的建筑形象。

建筑造型通过对鼎尊、风帆形态的创造性转换，令两种意向截然相反的元素自然结合，营造出"长风云帆、鼎立前海"的建筑意象。立面采用镂空肌理的超高性能混凝土挂板，既可以抵御海滨环境的侵蚀、遮阳通风，又增加了建筑的文化调性。

方案将建筑和基座整体向上抬升，在东侧和南侧形成良好的礼仪界面和主入口空间；建筑西侧和北侧则以台地及缓坡的方式与统筹设计区内的滨海景观带相接。首层架空层形成柱廊空间，结合下沉广场和外部礼仪广场整体设计成国深博物馆室外公共空间，即人群集散空间。地下设置停车库、藏品库房、交通转换空间、社会教育用房及文创休闲空间。地上展馆部分对传统"坊"及"九宫"的演绎转换，将原本两横两纵的均质巷道转变成以两纵为主，强调南北通透的连廊，呼应山海关系，形成新的空间模式。

The form of Guoshen museum is the integration of maturity with romantic and open characteristics, just shows the fusion of National Museum and the Qianhai Bay, forming a unique architectural form with multiple meanings.

Through the creative transformation of tripod and sail forms, the architectural form naturally combines the two elements which approaches in opposite intention, to make the building both modern and traditional, and creates an image of "as tripod, as sails, flourishes among the mountains and seas ". The facade adopts ultra-high performance concrete board with hollow texture to resist the erosion in the seaside environment, which not only increases the cultural meaning of the building, but also plays a role of shading and ventilation.

Vertically, the whole building is raised form a "platform", creating a good ceremonial interface and a main entrance space on the east and south sides. The west and north sides of the building are connected by terraces and gentle slopes to the coastal landscape. On the first floor, the colonnade space is open to the outside, to form a public space for gathering and distribution together with the sunken square and the exterior etiquette square. There are parking garage, collection warehouse, traffic conversion space, social education room and cultural and creative leisure space in the underground floors. Through the interpretation and transformation of traditional space layout "Fang" and "Nine palaces", the design transforms the original homogenous cross laneways of two horizontal and two vertical into two main vertical ways, emphasizing the north-south space corridor in site, echoing the relationship between mountains and seas, and forming a new spatial distribution framework for museums.

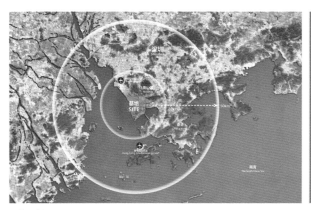 

区位分析 Location analysis

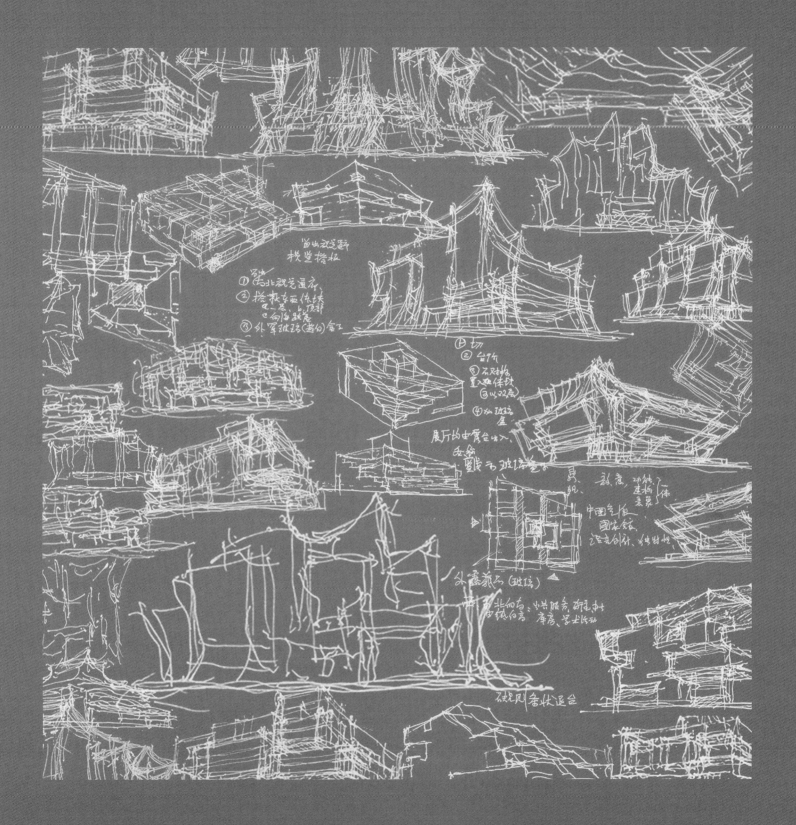

对页：西南向鸟瞰 Opposite: Aerial view from the southwest

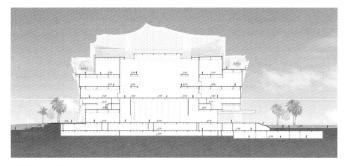

剖面图 Section

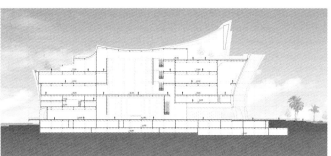

剖面图 Section

对页上图：主入口透视 Opposite top: Main entrance perspective
对页下图：礼仪广场透视 Opposite below: Etiquette square perspective

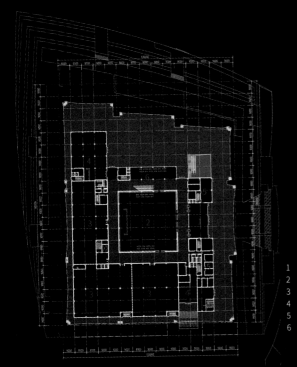

1	主门厅	Main lobby
2	礼仪大厅	Etiquette hall
3	陈列展厅	Exhibition hall
4	次门厅	Secondary entrance hall
5	附属用房	Subsidiary rooms
6	休息活动区	Rest area

一层平面图 Ground floor plan

1	中央大厅上空	Void above the central hall
2	陈列展厅	Exhibition hall
3	文物鉴赏室	Antiques estimation room
4	宴会厅	Banquet hall
5	准备间	Preparation room
6	接待区	Reception
7	通廊	Passage corridor
8	平台上空	Void above the terrace

六层平面图 Sixth floor plan

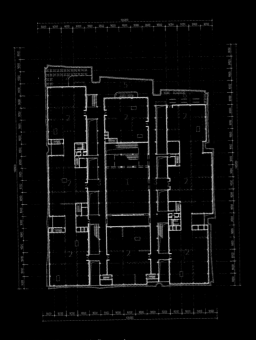

1	中央大厅	Central hall
2	陈列展厅	Exhibition hall
3	通廊	Passage corridor
4	休闲吧	Rest area
5	观海平台	Sea view desk
6	文创售卖区	Cultural and creative souvenirs

四层平面图 Fourth floor plan

1　屋顶花园　Roof garden

屋顶平面图 Roof floor plan

上图：观海平台 Top: Sea view deck
下图：山海通廊透视 Below: The mountain-sea corridor perspective
对页：中央大厅 Opposite: Central hall

北京城市副中心大剧院
建筑设计方案
GRAND THEATRE PROPOSAL OF BEIJING CITY SUB-CENTER

项目地点　中国·北京
项目规模　121200 m²
合 作 者[①]　王大鹏、刘翔华、张莹、王静辉、柴敬、杨涛
项目时间　2012 年设计，国际招标入围

Location　Beijing, China
Gross floor area　1304585.9 ft² (121200 m²)
Associates[①]　Wang Dapeng, Liu Xianghua, Zhang Ying, Wang Jinghui, Chai Jing, Yang Tao
Status　Designed in 2012, enter the finalist of international bidding.

对页：东南方向透视
Opposite: Southeast perspective

下图及对页：设计草图 Below and opposite: Design sketch

下图：设计草图 Below: Design sketch
对页：建筑模型 Opposite: Model

剧院以"京味水韵，写意通州；蓝绿交织，活力舞动；城市客厅，开放共享；根系传统，面向未来"为核心理念，以与行政中心平行的轴线为基准，沿运河展开，彰显大剧院的文化性和地标性的同时，充分与环境协调。四个剧场保持独立前厅，独立经营，但后台空间需要时可以共享。水绿景观与广场的活力人流通过剧场之间的公共通廊相互交织与渗透，天光自天窗洒下，人们在其中享受各类文化休闲活动，地下文化内街与剧院公共门厅、TOD相连，并与公共通廊、公共平台形成一个立体的空间体系，免费开放，触发城市活力。最后，连绵起伏的屋面覆于其上，京味水韵自然流露，融合古典与现代，与高台、柱廊一同传达东方智慧以及中华基因的魅力。

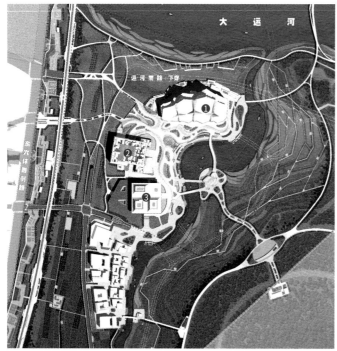

总平面图 General plan   0 50m

1 剧院   Theatre
2 图书馆 Library
3 博物馆 Museum

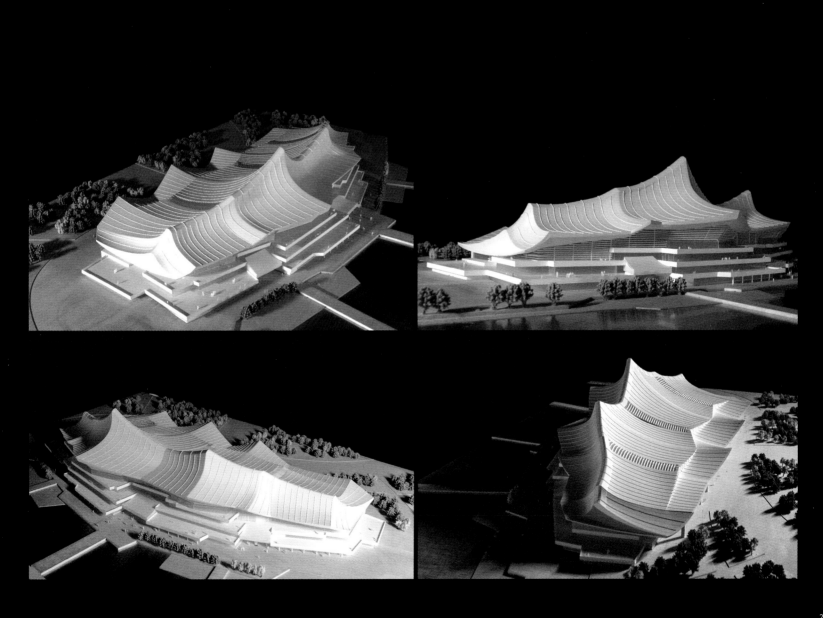

对页:东北方向鸟瞰 Opposite: Aerial view from the northeast

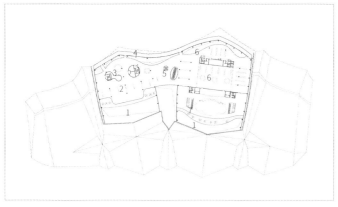

1	上空	Void
2	画廊	Gallery
3	服务区	Service area
4	观景走廊	View corridor
5	展览区	Exhibition area
6	文化阅览	Cultural exhibition and reading area
7	公共休息区	Public rest area

三层平面图 Third floor plan

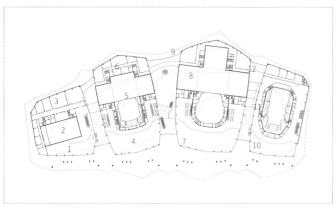

1	多功能厅门厅上空	Void above multi-functional lobby
2	多功能厅上空	Void above multi-functional room
3	多功能厅办公区	Offices of multi-functional room
4	戏剧院门厅上空	Void above lobby of theatre
5	戏剧院上空	Void above theatre
6	戏剧院办公区	Offices of theatre
7	歌剧院门厅上空	Void above lobby of opera house
8	歌剧院及舞台上空	Void above stage
9	歌剧院办公区	Offices of opera house
10	音乐厅门厅上空	Void above lobby of concert hall
11	音乐厅上空	Void above concert hall
12	音乐厅办公区	Offices of concert hall

二层平面图 Second floor plan

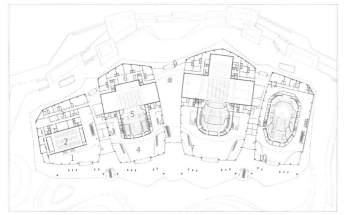

1	多功能厅门厅	Multi-functional lobby
2	多功能厅	Multi-functional room
3	多功能厅办公区	Offices of multi-functional room
4	戏剧院门厅	Lobby of theatre
5	戏剧院	Theatre
6	戏剧院办公区	Offices of theatre
7	歌剧院门厅	Lobby of opera house
8	歌剧院及舞台	Opera house and stage
9	歌剧院办公区	Offices of opera house
10	音乐厅门厅	Lobby of concert hall
11	音乐厅	Concert hall
12	音乐厅办公区	Offices of concert hall

一层平面图 Ground floor plan

对页：西南方向半鸟瞰 Opposite: Aerial view from the southwest

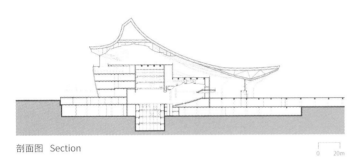

剖面图 Section 0 20m

"To combine the traditional Beijing tone with the water, to twist the water and the green belt, to build a city hall open for the whole, to absorb from the past and face to the future", we come up with this design. The axis is paralleled to the administrative center, the grand theatre is developed along the canal, highlighting the cultural and landmark nature of the grand theater while fully coordinating with the environment. The four theatres maintain separate front rooms and operate independently, but the backstage space can be shared when needed. Aqua green landscape and energetic flows of people through the theatre of the square are intertwined and infiltration between the public corridors, the skylight to sprinkle, in which people enjoy all kinds of cultural and leisure activities, underground culture street and theatre communal hallway, TOD is linked together, and with the public corridor, the public platform to form a three-dimensional space of system, opening to the outside world, stimulate vitality city. Classical and modern fusion, with the bottom volume, colonnade together to convey the wisdom of the east, the charm of the Chinese gene.

下图：西面鸟瞰图 Below: Aerial view from the west
对页：东面鸟瞰图 Opposite: Aerial view from the east

注释

① 本次招标方案包括城市设计及大剧院、图书馆、博物馆三个单体建筑。由东南大学建筑设计研究院有限公司、深圳市建筑设计研究总院有限公司、北京土人城市规划设计股份有限公司、筑境设计四个单位，采用工作营方式完成该设计。本书仅列入由筑境设计负责的大剧院部分。

① The bidding proposal includes the urban design and building design of the grand theater, library, museum and other three single buildings. The design was completed by a workshop among the four cooperators: the Architects & Engineers Co.,LTD. of Southeast University, the Shenzhen Architecture Design Co.,LTD, TURENSCAPE and CCTN. Only the grand theater design under the charge of CCTN is included in this book.

下图：大剧院前厅透视 Below: Theatre lobby perspective
对页上图：文化休闲空间透视 Opposite top: Cultural leisure space perspective
对页中图：休闲区室内透视图 Opposite middle: Indoor leisure space perspective
对页下图：棕榈园透视图 Opposite Below: Palmengarten perspective

剧院顶部的文化展示区及休闲餐饮区直面运河，具有独一无二的景观视角。在这里，市民可以拥抱运河，品味文化，沉浸在艺术的殿堂之中。

The cultural exhibition area and the casual dining area above the theatre, facing the canal, have a unique view. Citizens can embrace the canal, taste culture and immerse themselves in the palace of art here.

中华国乐中心·江阴
CHINESE NATIONAL MUSIC CENTER, JIANGYIN

项目地点　中国·江苏·江阴
项目规模　108340 m²
建筑主创　程泰宁、张莹
合 作 者　殷建栋、吴妮娜、张莹、闵杰、何其乐、周羽歆、谢晓峰、
　　　　　　陈志浩、乔琪、高良杰、卓御杰、郑佳晟、刘文雯
项目时间　2017年设计，扩初设计阶段

Location　Jiangyin, Jiangsu Province, China
Gross floor area　1166162.1 ft² (108340 m²)
Chief architect: Cheng Taining, Zhang Ying
Associates　Yin Jian dong, Wu Nina, Zhang Ying, Min Jie, He Qile,
　　　　　　Zhou Yuxin, Xie Xiaofeng, Chen Zhihao, Qiao Qi,
　　　　　　Gao Liangjie, Zhuo Yujie, Zheng Jiasheng, Liu Wenwen
Status　Designed in 2017, under design developing.

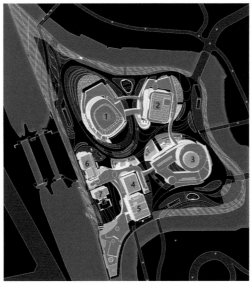

总平面 Site Plan

对页：鸟瞰图
Opposite: Aerial view

下图：设计草图 Below: Design sketch
对页：建筑模型 Opposite: Models

设计立足江阴深厚的国乐基础，以北门岛的开发作为平台，致力发展以中华国乐为主题的文化产业，打造激活江阴、服务江苏、辐射全国、走向世界的"中华国乐中心"。

Based on the profound foundation of Chinese music in Jiangyin and taking the development of Beimen Island as the platform, the design is committed to developing the cultural industry with the theme of Chinese national music, and creating a "Chinese national music center" that activates Jiangyin, serves Jiangsu, radiates to the whole country and goes to the world.

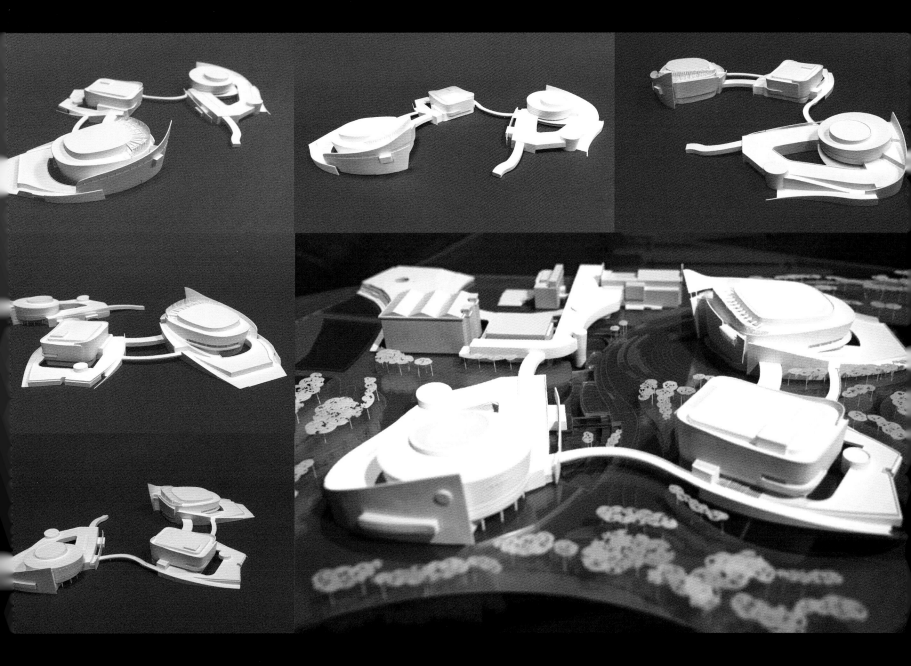

下图：音乐厅透视 Below: Concert hall perspective

北门岛区位独特，坐拥锡澄运河，毗邻长江、君山、北接滨江商务片区，东临城市老城区，西临城市新组团，正处江阴城市东西发展轴和南北运河绿轴的交汇之处，是江阴未来城市发展的中心区域。

立足于"国乐之岛，城市之芯"特殊定位，综合考虑文化与功能诉求，通过多元复合的总体布局、丰富开放的公共空间设计，以及行云流水的建筑形象。将北门岛打造成为激发城市活力的文化中心和江阴的城市文化新地标。

Beimen Island has a unique location, with Xicheng canal alongside, adjacent to the Yangtze River and Junshan mountain, and having Binjiang business district to the north, the old city to the east, and the new city group to the west. It is located at the intersection of the east-west development axis of Jiangyin city and the green axis of the north-south canal, and is the central area of future urban development of Jiangyin.

Based on the unique positioning of "An island of national music, the heart of the city", the comprehensive consideration of cultural and functional appeals, through the manipulation of general layout, a public space of openness and richness is designed, as well as the flowing architectural form. Beimen Island will be built into a cultural center that stimulates the vitality of the city and a new cultural landmark of Jiangyin.

对页：科技馆、音乐厅沿河透视 Opposite: Riverside perspective of the science and technology museum and the concert hall

1	屋顶平台	Roof deck	1 音乐厅门厅	Lobby of concert hall
2	音乐厅上空	Void above concert hall	2 音乐厅	Concert hall
3	音乐厅休闲区	Restspace of concert hall	3 室外庭院	Outdoor garden
4	美术馆展厅	Exhibition hall of gallery	4 国乐体验展区	National Music Experiencing Area
5	空中连廊	Air corridor	5 美术馆门厅	Lobby of gallery
6	科技馆螺旋坡道	Spiral ramp of science and technology museum	6 美术馆中庭	Courtyard of gallery
7	科技馆展厅	Exhibition hall of science and technology museum	7 美术馆拍卖厅	Auction house of gallery
8	妇幼活动中心活动室	Activity Room of Women and Children Activity Center	8 妇幼活动中心中庭	Courtyard of Women and Children Activity Center
9	营地中心活动室	Activity Room of Camp Center	9 科技馆中庭	Courtyard of science and technology museum
10	妇幼活动中心活动区	Activity area of Women and Children Activity Center	10 科技馆营地中心	Camp center of science and technology museum
11	公共活动区	Public event area	11 公共架空层	Public stilt floor
12	配套食堂（改造）	Canteen (transformed)	12 配套食堂（改造）	Canteen (transformed)
13	多功能厅上空（改造）	Void above multifunctional room (transformed)	13 多功能厅（改造）	Multifunctional room (transformed)
14	青少年宫二层（改造）	Second floor of Children's Palace (transformed)	14 管理用房（改造）	Offices (transformed)
15	展厅（改造）	Exhibition hall (transformed)	15 青少年宫（改造）	Children's Palace (transformed)

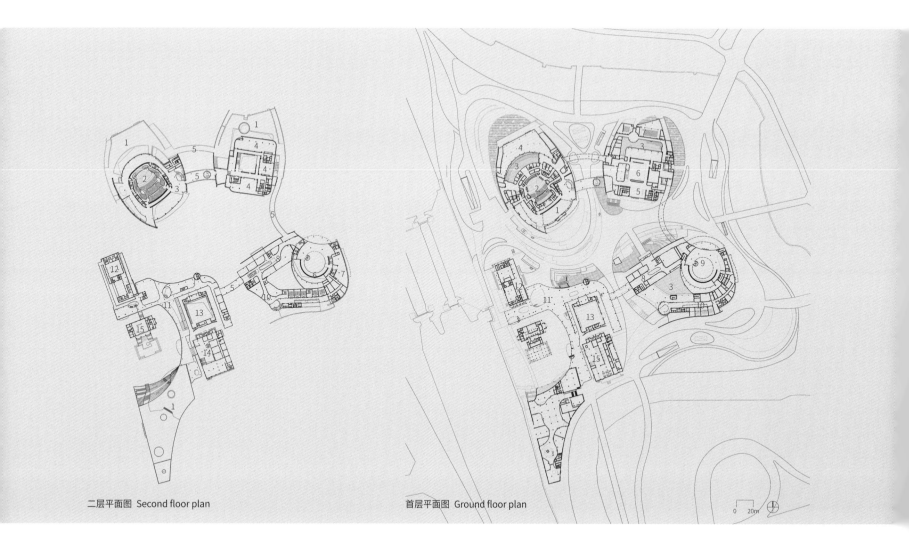

二层平面图 Second floor plan

首层平面图 Ground floor plan

下图：科技馆透视 Below: Science and technology museum perspective
对页：全岛半鸟瞰 Opposite: Aerial view of the island

1	展厅	Exhibition Hall	10 临时展厅	Temporary Exhibition Hall
2	机房	Computer Center	11 艺术品拍卖厅	Art Auction House
3	中庭	Central Courtyard	12 内院	Inner Garden
4	非机动车库	Non-motorized Garage	13 休闲大厅	Free Hall
5	5d 影院	5D Cinema	14 配套服务区	Equipment and Services
6	办公区	Office	15 表演休息区	Performance Lounge
7	回廊	Cloister	16 国乐音乐厅	National Music Hall
8	门厅	Lobby	17 入口门厅	Entrance Hall
9	阅览区	Reading Area		

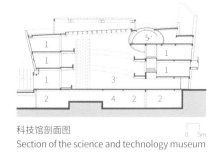

科技馆剖面图
Section of the science and technology museum

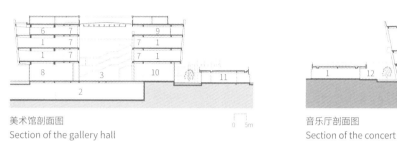
美术馆剖面图
Section of the gallery hall

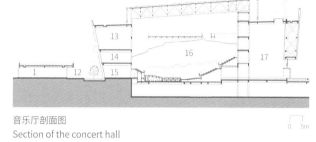
音乐厅剖面图
Section of the concert hall

南京雨花华邑酒店
NANJING YUHUA HUAYI HOTEL

项目地点　中国·江苏·南京
项目规模　45627 m²
合作单位　江苏省建筑设计研究院有限公司
　　　　　景观设计：TOPOS 拓柏景观
　　　　　室内设计：深圳杨邦胜室内设计有限公司
合 作 者　殷建栋、吴妮娜、杨涛、鲍张丰、沙吟、陈忠鹤
项目时间　2019 年设计，扩初设计阶段，施工中

Location　Nanjing, Jiangsu Province, China
Gross floor area　491124.94 ft² (45627 m²)
Cooperators　Jiangsu Provincial Architectural D&R Institute LTD
　　　　　　　Landscape Design: TOPOS Landscape Architects
　　　　　　　Interior Design: YANG & ASSOCIATES GROUP
Associates　Yin Jiandong, Wu Nina, Yang Tao, Bao Zhangfeng, Sha Yin, Chen Zhonghe
Status　Designed in 2019, under design developing.

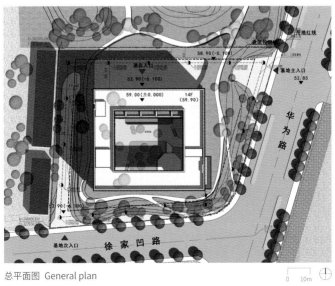

总平面图　General plan

对页：东南向透视
Opposite: Southeast perspective

南京雨花华邑酒店 NANJING YUHUA HUAYI HOTEL

下图：设计草图　Below: Design sketch
对页：鸟瞰图　Opposite: Aerial view

南京雨花华邑酒店项目位于中国（南京）软件谷范围内。中国南京软件谷位于南京市雨花台区，是中国第一软件产业基地，中国最大的通讯软件产业研发基地，国家重要软件产业和信息产业中心。基地坐标华威钢构以东，华为路以西，徐家凹路以北，紧邻江苏科技大厦、华威钢构、华为云楼等软件谷已建成建筑。

Nanjing Yuhua Huayi Hotel is located within the scope of China (Nanjing) software Valley. Nanjing Software Valley, located in Yuhuatai District, Nanjing city, is China's first software industry base, China's largest communication software industry research and development base, national important software industry and information industry center. The coordinates of the site are east of Huawei Structure Company, west of Huawei Road, north of Xujiaao Road, close to Jiangsu Science and Technology Building, and some buildings in the software valley such as Huawei Steel Structure Buildling and Huawei Cloud Building.

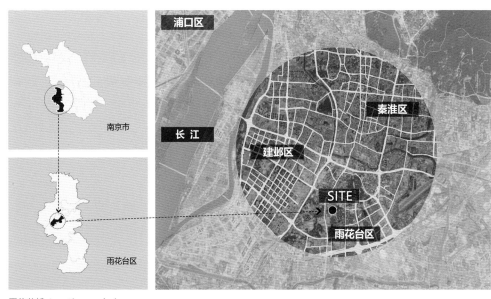

区位分析 Location analysis

1	宴会门厅	Hall lobby	9	新娘房	Bride room
2	酒店前厅	Hotel lobby	10	台地景观	Terrace landscape
3	宴会厨房	Kitchen	11	大堂	Lobby
4	内部员工培训教室	Staff training room	12	餐厅	Restaurant
5	卫生间	Rest room	13	露台花园	Terrace garden
6	宴会厅	Hall	14	大堂吧	Lounge
7	服务廊兼备餐	Food preparation & service	15	茶室	Teahouse
8	前厅	Lobby			

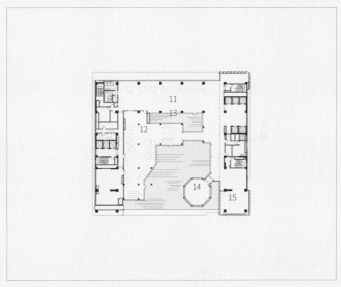

三层平面图 Third floor plan

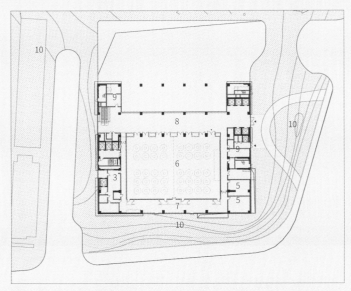

一层平面图 Ground floor plan

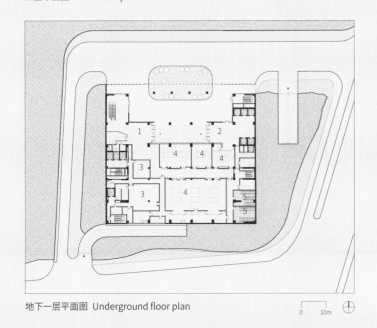

地下一层平面图 Underground floor plan

0 10m

下图：东南街角局部透视 Below: Southeast perspective at the street corner
对页：西南向透视 Opposite: Southwest perspective

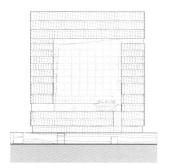
南立面图 South elevation

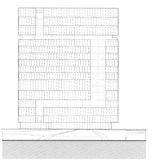
北立面图 North elevation

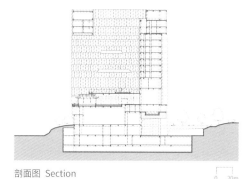
剖面图 Section

上图：入口透视 Top: Entrance perspective
下图：酒店大堂入口 Below: Entrance of the hotel lobby
对页：庭院空间 Opposite: Courtyard

拟建项目位于南京市雨花台砂砾石层保护区南侧，该保护区属于国家环境保护区名录之一。项目着力考虑对保护区的影响。整体设计意图与北侧地貌一体化，打造城市中的山地地貌。建筑形体取义《道德经》之"无极"，表达对古代哲学思想的抽象理解，体悟无中心、无边际之意。设计策略围绕"恢复山地地形、保留城市记忆，公园最大化，作为公共空间向城市开放，整体的建筑形态"四点展开。

南京雨花华邑酒店项目的功能定位为五星级酒店，采用洲际华邑酒店品牌，构建集住宿、会议、餐饮、康娱、体验于一体的全服务型酒店。

The designed project is located in the south of the Yuhuatai Gravel Bed Protection Zone in Nanjing, which is one of the national environmental protection areas. The design focuses on the impact on the reserved zone. The design intention is to integrate the building with the northern landform to create the mountainous land image in the city. The architectural form takes the meaning of "the infinite" in *Tao Te Ching* to express the abstract understanding of ancient philosophical thoughts and realize the meaning of no center and no margin. The design strategy focuses on four aspects:"restoring mountainous terrain and preserving urban memory, maximizing city parks, opening to the city as a public space, creating a united architectural form".

Nanjing Yuhua Huayi Hotel aims at five-star hotel, using intercontinental Huayi hotel brand to build a full-service hotel integrating accommodation, conference, catering, entertainment and experiencing.

左图、中图：酒店大堂 Left and middle: Hotel lobby
右图：多功能厅 Right: Multi-function room

徐州园博会宕口酒店
DOWN HOLE HOTEL OF XUZHOU GARDEN EXPO

项目地点　中国·江苏·徐州
项目规模　29597 m²
合 作 者　王大鹏、蓝楚雄、盛思源、张潇羽、邱培昕、廖慧雯、
　　　　　肖华杰、施妙佳
项目时间　2020 年设计，施工中

Location　Xuzhou, Jiangsu Province, China
Gross floor area　318579.46 ft² (29597 m²)
Associates　Wang Dapeng, Lan Chuxiong, Sheng Siyuan,
　　　　　Zhang Xiaoyu, Qiu Peixin, Liao Huiwen, Xiao Huajie,
　　　　　Shi Miaojia
Status　Designed in 2020, under construction.

　　本项目基地处于徐州市铜山区园博园区西南角的综合展馆区，场地内主要是采石开挖的宕口，地形起伏较大。宕口酒店的区位远离城市中心，综合考虑酒店其度假性质的定位，遵循"绿色、创新"的原则，结合建筑与宕口之间的联系，在最大程度利用山体景观资源的同时，为游客提供独特的入住体验。

对页：西北向鸟瞰
Opposite: Northwest aerial view

下图及对页：设计草图 Below and opposite: Design sketch

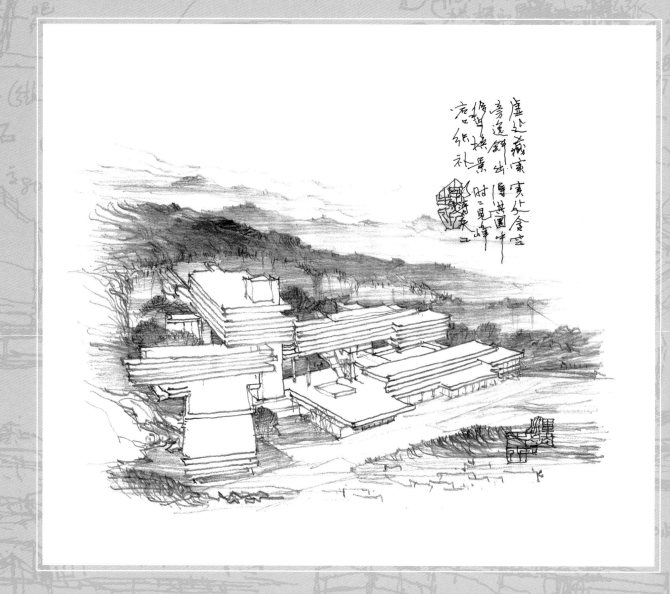

对页：东北向日景透视 Opposite: Northeast perspective

区位分析 Location analysis

The site of this project is located in the comprehensive exhibition area at the southwest corner of Garden EXPO Park, Tongshan district, Xuzhou city. The site is mainly occupied by the openings of stone excavation, and the terrain fluctuates greatly. The location of Dangkou Hotel is far away from the city center. Considering the positioning of resort nature, Dangkou Hotel follows the principle of "green and innovation" and combines the connection between buildings and openings of stone excavation to make the best use of mountain landscape resources and provide unique stay experience for tourists.

对页：东北向夜景透视 Opposite: Northeast perspective at night

设计理念——依山观湖，织补宕口，悬空建筑，立体园林

依山观湖：本方案背靠宕口生长，面向悬水湖，拥有开阔的景观视野；

织补宕口：设计考虑到宕口景观修复，将其修复后的宕口景观通过建筑的局部架空与错层渗透出来；

悬空建筑：设计采用了错落有致的建筑形体，在不同的标高上设置建筑主体；

立体园林：最大程度利用山体景观资源生成立体复合的园林空间。

Design concept - view the lake by the mountain, stitching the openings of stone excavation, hanging buildings, three-dimensional gardens.

View the lake by the mountain: The hotel is backed by openings on mountain and facing the Xuanshui Lake, with an open landscape view.

Stitching the openings of stone excavation: The design takes the restoration of landscape into consideration and the restored landscape is permeated through building by open floor and terrain-like form of the building.

Hanging buildings: the building is divided into parts to hide itself into the mountain in different levels.

Three-dimensional gardens: make full use of mountain landscape resources to generate three-dimensional complex garden space.

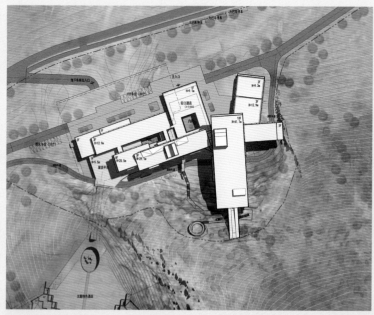

总平面图 General plan

对页：主入口日景透视图 Opposite: Main entrance perspective

1 酒店大堂 Lobby	4 风味餐厅 Restaurant	7 客房区 Rooms
2 大堂吧 Lounge	5 自助餐厅 Buffet	8 露台 Terrace
3 会见室 Meeting room	6 辅助用房 Subsidiary rooms	9 咖啡厅 Coffee bar

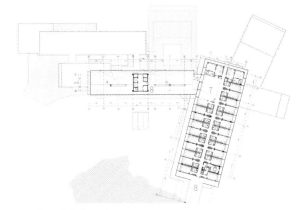

八层平面图 Eighth floor plan

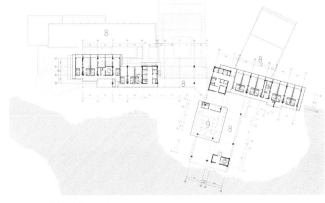

五层平面图 Fifth floor plan

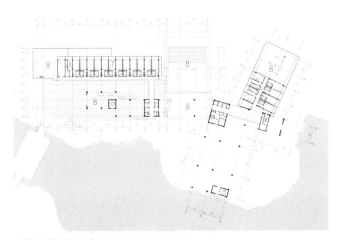

三层平面图 Third floor plan

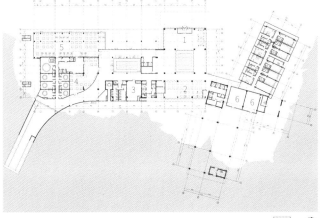

一层平面图 Ground floor plan

0　　10m

本图为施工基本完成后拍摄,为实景。
This picture is taken after the construction and it is basically completed which shows a real scene.

1 客房　Rooms
2 露台　Terrace
3 咖啡厅　Coffee bar
4 大堂吧　Lounge
5 大堂　Lobby
6 地下车库　Underground parking garage

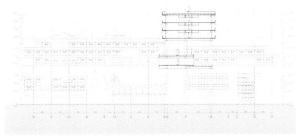

剖面图 Section

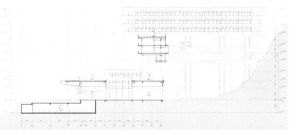

剖面图 Section

下图及对页：局部透视 Below and Opposite: Perspectives

上图：大堂吧 Top: Lounge
下图：大堂 Below: Lobby
对页上图：多功能厅 Opposite top: Multi-function room
对页下图：连接通道 Opposite Below: Connecting passage

中国水工科技馆
CHINA WATER CONSERVANCY ENGINEERING SCIENCE AND TECHNOLOGY MUSEUM

项目地点　中国·江苏·淮安
项目规模　107500 m²
合 作 者　殷建栋、古振强、邱培昕、闵杰、陈阳、牟旭楠、周羽歆、
　　　　　郭义豪
项目时间　2019 年设计，施工图深化中

Location　Huai'an, Jiangsu Province, China
Gross floor area　1157120.4 ft² (107500 m²)
Associates　Yin Jiandong, Gu Zhenqiang, Qiu Peixin, Min Jie,
　　　　　　　Chen Yang, Mu Xunan, Zhou Yuxin, Guo Yihao
Status　Designed in 2019, construction drawings in progress.

　　项目以中国水工科技馆建设为核心，串接四馆一院，整合板闸遗址公园，其中水工馆、会议中心与板闸遗址公园呈品字形布局，围合形成秋月广场，靠近里运河则引水入园塑造户外水工展园，打造集遗址保护、水工科技展示、户外水上体验等于一体的，多元复合的城市文化建筑集群。

The design of China Water Conservancy Engineering Science and Technology Museum takes the construction of Water Conservancy Engineering Museum as the core, connects four museums and one courtyard in series, and integrates the gate ruins park. Among them, the Water Conservancy Engineering Museum, the Conference Center and the Banzha Relics Park are arranged in a shape, forming the Autumn Moon Square. Water will be brought into the garden near the canal to shape an outdoor water conservancy exhibition garden, and create a multi-functional urban cultural complex that integrates heritage protection, water conservancy science and technology display and outdoor water experience.

对页：沿河透视
Opposite: Perspective along the river

下图：设计草图 Below: Design sketch

闸口过漕图（淮安清江大闸历史复原图）
Canal transport boat passing through the water-gate (restored map of Qingjiang River Grand Gate in Huai'an)

板闸遗址 Site of flat gate

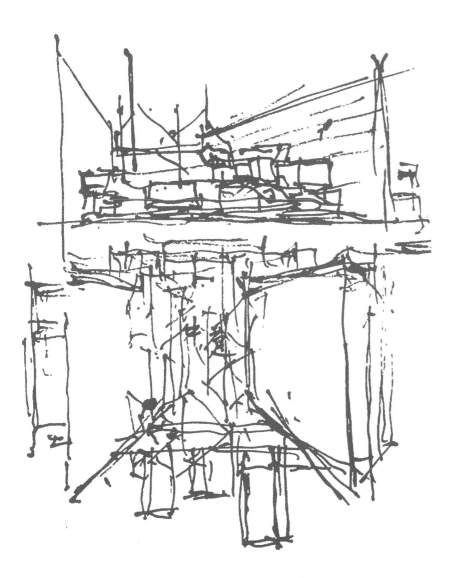

水工馆的建筑设计基于"文化性、现代性、公共性"的设计原则，从表现有淮安特色的水工文化和漕运文化出发，以"千舟竞渡 运河扬帆"为概念意向，大实大虚的体量关系兼具雕塑感与功能性。两侧实体作为展厅功能，形态上呼应板闸遗址，面向运河打开，表现出拥抱运河之势；中部虚体作为公共空间，取意大运河河水，串联起运河与城市，塑造开放共享的城市形象；穿插其中的小尺度体量犹如漕船过闸奔涌而出，于运河中扬帆起航，极具动势，同时作为兼有休闲观景功能的亲水平台，室内外一体化设计加强展馆与运河的互动与关联。

　　中国水工科技馆的设计是对于中国水工文化的当代诠释，将助力打造根植于淮安地域文化和大运河文化的以水工为主题的科技展馆。

The architectural design of the Water Conservancy Engineering Science and Technology Museum is based on the design principle of "culture, modernity and publicity". Starting from the expression of the water conservancy engineering culture and the canal transport culture with Huaian characteristics, the concept intention of "Thousands of boats racing and sailing in the canal" is taken as the conceptual intention. The relationship between building volumes is both sculptural and functional. The volumes on both sides serve as the functions of the exhibition hall, echoing the sluice site in shape, opening towards the canal to show the trend of embracing the canal; The central open space as a public space, taking the meaning of the Grand Canal water, connecting the canal and the city, to create an open and shared city image; The small volume is like a water canal boat rushing out of the lock and setting sail in the canal, which is highly dynamic. At the same time, as a horizontal platform with leisure viewing function, the indoor and outdoor integrated design strengthens the interaction and association between the exhibition hall and the canal.

The design of China Water Conservancy Engineering Science and Technology Museum is a contemporary interpretation of China's water engineering culture, and will help build a science and technology exhibition hall with the theme of water engineering rooted in the regional culture of Huai'an city and the Grand Canal.

区位分析 Location analysis

对页：西侧鸟瞰 Opposite: Aerial view from the west

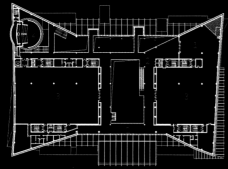

1　侧厅 / 多功能厅　　Multi-function hall
2　内廊　　　　　　　Corridor
3　主展厅 / 多功能厅　Exhibition hall/multi-function hall
4　飞翔影院　　　　　Theatre

三层平面图 Third floor plan

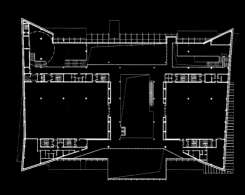

1　开放临时展区　　　Emporary exhibition area
2　内廊　　　　　　　Corridor
3　主展厅 / 多功能厅　Exhibition hall/multi-function hall

二层平面图 Second floor plan

1　入口门厅　　　　　Lobby
2　中庭　　　　　　　Courtyard
3　主展厅 / 多功能厅　Exhibition hall/multi-function hall
4　影视区　　　　　　Movie space
5　青创 / 餐饮　　　　Sale/restaurant
6　纪念品商店　　　　Souvenirs
7　办公区　　　　　　Offices
8　培训区　　　　　　Training rooms

一层平面图 Ground floor plan

下图：设计草图 Below: Design sketch
对页：水广场透视 Opposite: Water square perspective

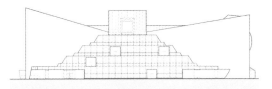
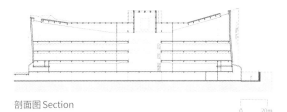

Q 轴 -A 轴立面图 Axis Q-A elevation　　20 轴 -1 轴立面图 Axis 20-1 elevation　　剖面图 Section

上图：立体栈道　Top: The three-dimensional road
下图：水岸横厅　Below: Water bank hall
对页：中央大厅　Opposite: Central hall

学术访谈　ACADEMIC INTERVIEW

语言·意境·境界
—— 程泰宁院士建筑思想访谈录

费移山：费　　程泰宁：程

原载于《建筑学报》2018年10期

2017年初，我开始协助程泰宁院士筹备《境界与语言》一书的写作工作。最初我曾计划通过一两次的问答，对程泰宁院士的建筑思想建立一个通盘认识。熟料在此之后，这样的采访经历了有数十次之多。每一次采访都犹如小径分叉的花园，思路纷繁，且不少观点与主流建筑学颇不相同，常使我在最初产生困惑、茫然之感。而随着讨论的逐步深入，我渐渐对这些思想线索及它们与当代建筑学之间的关系产生了理解，有了一些自己的体悟。

正如程泰宁院士自己所言："建筑学研究最关键的就是'化'，要融通中外，转换提升，化入心中。"而对于程泰宁院士来说，他所有思考又都是建立在实践土壤之上的，是实践与思考的彼此参照，相互推进，因此也就具有了与一般理论研究不同的价值与意义。讨论中，程泰宁院士多次提醒，建筑学具有复杂性、多义性与开放性特征，是很难用条分缕析式的简单理性来梳理清楚的。所以，对程泰宁院士思想网络的最好呈现，也许就是他本人的建筑作品与那些转瞬即逝的话语本身。希望现在为了行文之便所做的权宜之计，没有贬损那些话语本身的思想价值。

本文是过去两年中，数十次讨论内容的筛选与整理。其中前半部分主要是对当前建筑设计领域发展方向的思考；后半部分则是对程泰宁院士所提出的"境界、意境、语言"这一建筑思想体系的诠释。

对于中国建筑发展方向的思考

费：在大家的印象中，您是一位以实践见长的建筑师，完成过许多让人印象深刻的建筑作品。但是近几年，您似乎突然开始关心理论研究了，写了好几本书，并更多地在学术界发声。这种转变背后的原因是什么呢？

程：作为建筑师，一直以来我都保持着边创作、边思考的工作习惯。过去每完成一个重要的建筑作品，或者在创作实践中有所心得，都会写一些理论文章寄给《建筑学报》发表。对我来说，实践与理论总是相辅相成的，这中间并不存在一个突然转变。但是这两年，我的确比之过去更加关注理论研究。这主要是因为现在建筑设计及其相关领域中的矛盾特别突出，要解决这些问题，理论研究就显得特别重要。同时也是因为经过这么多年的积累，我自己对建筑创作中的许多问题有了更为清晰的认识，到了可以深入谈一谈的时候了。

费：您所说的这个"矛盾"具体指的是什么呢？

程：我认为，困扰当前中国建筑设计领域的最根本矛盾仍然是价值观的问题。什么是好的建筑？我们已经失去了明确的判断。

这其中，建筑的异化是非常重要的一个方面。好的建筑应该能满足适用、安全、生态节能以及技术经济合理等基本要求，也就是国际建协《北京宪章》所说的"回归基本原理"。但是在当下，很多建筑正在偏离其本应该承载的社会功能与文化价值，而被异化为某些人夸功炫富的宣传工具，成为一种被消费、被娱乐的商品。在这样一种趋势的影响下，使得我们很难用正常的价值标准来评价建筑，也在很大程度上导致建筑设计思想陷入混乱。

另外一个重要方面，是面对西方的时候，文化上的自我矮化。我始终认为，西方建筑对中国建筑发展正反两个方面的影响都不能被低估。这两年谈文化自信的人多了起来，但是无论是在社会上，还是我们自己领域内部，"月亮还是外国圆"的思想仍然根深蒂固。我们不少领导和开发商仍然在崇洋求怪；大型项目设计招标，中国建筑师仍然需要"傍"上老外，最近南方某特大城市的一个文化项目进行招标，仍然只允许国外建筑师参加。而在建筑设计领域内部，以"西方"为"现代"，以抄袭模仿为"接轨"，也是近30年来的一个普遍现象。这种一方面大谈文化自信，另一方面却被西方裹挟的现象，不就是当下矛盾的一种突出体现吗？

我认为不解决这些矛盾，就不可能实现中国建筑设计水平的总体提升。

费：我能理解您对现在盲目崇洋现象的担忧，但同时我也有一些疑虑。比如说您多次谈到的中西之间价值观的不同。这对于中国建筑未来发展的影响体现在哪里？建筑设计是否可以按照中、西来划分呢？

程：大家可能会有一个误解，认为我谈"中国的建筑"仅仅是出于一种民族自尊心，是故意要与西方建筑唱反调。其实完全不是这样。我自己的建筑学教育就是从西方的布扎体系开始的。之后我和大家一起经历了现代主义、后现代主义、建构、参数化等各种理论思潮与方法体系的更新。年轻时有很长一段时间，我对西方建筑很痴迷，在20世纪80年代我还写过呼吁加强中西方交流的文章。说我不受这些西方建筑理论与方法的影响，那是不可能的。但正因为喜欢、关注，我才既知道它们的好，也逐渐认识到它们的局限性。

我认为对西方建筑必须作历史的、全面的观察，不能为一个时期、一种流派所局限。就拿当下来说，西方建筑的发展所面临的问题与矛盾很多。前一阵，日本著名建筑师槙文彦在一次访谈中说了一段话，大意是20世纪50年代的现代主义是一艘人人都想搭乘的大船，如今大船没有了，只有无数漂浮在大海上的碎片。作为同代人，我也有同感。随着西方进入后工业社会以后，在后现代文化的冲击之下，西方建筑呈现出碎片化、反理性的发展倾向。在这样一种情况下，如果我们再把大海之中漂浮着的某个碎片当成自己的方向，就太不可取了。所以我谈中国建筑的价值建构，不是故意要和西方建筑唱对台戏，而是去探索自己的方向，这也是摆脱漂浮和跟风的唯一办法。

我想说，中西方文化，包括建筑文化都正处于发展的某一个节点上。多元探索，文化重构，是这个时代的特点，对此我们要有充分的认识。

费：对于中国建筑的未来发展，现在不少建筑师与建筑学者可能会有不同的看法，比如说他们可能更多的是关注自己领域内部的探索，而不太关心哪些是属于"中国"的，哪些是属于"西方"的。您对此是怎么看的？

程：在我看来，中、西之间的差异肯定是萦绕在不少中国建筑师心头的难解之结，双方政治、经济、文化的碰撞是客观存在的。如何理解中国与西方之间的关系，是很多中国建筑师几乎每天都会面对的问题。而对这个问题的深入思考，就构成了中国建筑理论中重要的一部分。

那么为什么有人会对探讨中西文化之间的这种不同"不感兴趣"呢？我想这里面的原因可能有两个。

一是因为中国的建筑师长期处于西方主义的影响之下。近代以降，每个中国人的知识体系、思维方式都烙下了深刻的西方痕迹。而就建筑学来说，我们从学习建筑的第一天开始，受的就是一套来自西方的建筑教育。正是因为长期身处其中，所以往往会"直把他乡做故乡"。

除此之外，另一个更为重要的原因可能是这个问题太大，客观因素很多，很难说清楚。所以现在有不少建筑学者、建筑师对这些话题刻意回避。他们不是对这些问题没有感受，但是可能觉得这个话题过于宏大，不是建筑学、建筑设计所能够解决的，很多问题说了也白说，说了也说不清楚，所以不愿意多谈。

对于前一点，需要建筑师特别是年轻建筑师独立思考，假以时日，他们

1. 古巴吉隆坡胜利纪念碑国际设计竞赛参赛方案
International Competition for Monument of Pig-Bay, Cuba

2. 河姆渡遗址博物馆竞赛方案
Competition Design He-mu-du Relics Museum, Yuyao

3. 杭州黄龙饭店
Dragon Hotel, Hangzhou

自然会明白，中西建筑文化的区别究竟何在？而对后一种看法，我虽然能理解，但却不能接受。既然问题已经客观存在，我们就必须要认真地去面对它。不能因为这个问题很大、很难，我们就刻意回避，不去讨论。正因为大，因为难，才需要大家从自己的知识领域出发，一点点去努力，积小胜为大胜，最终才能有所突破。我们常说，世界应该向着多元化的方向发展。我们应该思考一下，中国建筑师与建筑学者能为这个多元化的世界贡献一些什么？这不算是过高要求吧。

费：那么您近来所强调的理论研究能对当下中国建筑的发展起到怎样一种作用呢？或者说，您是怎么看待当下国内理论研究与建筑创作实践之间的关系？

程：对于当下的中国建筑来说，理论研究的作用的确非常重要。你可以回想一下，近百年来，中国建筑师不是不想做一些自己的东西，例如20世纪30年代就曾经提倡过"中国固有之形式"；20世纪50年代倡导过"民族风格"；20世纪80年代"夺回古都风貌"也一度是当时建筑界的热点话题；近年来也不乏对"新而中""新中式"这样一些话题的讨论。但是由于缺乏有力的理论体系作为支撑，只是以形式语言反形式语言，以民粹主义排斥外来文化，热闹一阵之后就被迅速淹没在西方化的影响之下了。这说明没有理论支撑，仅靠民族自豪感，靠"跟风"，是不可能实现中国建筑的真正发展。在我看来，实践创新、理论思考与价值体系建构，是相辅相成的。实践层面所反映出来的许多问题都与评价体系相关，而这归根结底还必须依靠理论建构去解决。没有了理论支撑，文化自信就成了一句口号，而实践创新也不可能真正实现。

中国建筑需要从观念上、本体上去建构一些东西，形成中国建筑的思想与理论内核，并逐步建立自己的建筑理论体系。世界建筑就如一片多姿多彩的森林，中国建筑也是其中的一个"科目"。我们不能只是满足于在西方建筑的那棵大树上嫁接一个小枝条。无论是海南的椰树，还是西北的胡杨，它们都是从自己的土地上独立生长起来的。这几年我参与了一些理论研究的课题，主持完成了工程院课题——"当代中国建筑设计现状与发展"，接受了一些访谈，也写了一些文章，除了希望自己能为中国建筑的发展方向探探路以外，更希望通过这些工作，能引起大家对这些问题的重视，让更多的人投入到这个领域的工作中来。

费：现在还是有不少中国的建筑学者、建筑师在切实地关注中国"自己"的问题，您对这些研究是怎么看的？

程：近些年来，有不少学者和建筑师在一些对具体问题的研究方面做了扎实的工作，比如说对材料的应用和传统工艺的继承与发展等，也出现了不少颇有特点的建筑作品。在建筑理论研究，包括建筑创作实践中，从一些具体问题切入是一个办法，比较容易为人所理解，也比较好入手。我自己对建筑创作理论的探索也是从具体问题开始的。这些对具体问题的研究，可以看作是讨论一些更大问题之前的积累。没有这些研究，我们对中国建筑理论的讨论就没有了立足点。

但我们也要看到，当下对中国建筑的研究还缺乏系统性与深度。很多研究者总是自动地将"中国"建筑理论看作是对当下建筑学的一种"补充"。所谓"补充"那就是可深可浅、不追求系统性。所以很多理论探索总是浅尝辄止，不求体系。正是这样的观点大大局限了中国建筑理论的发展。

从我个人的创作经历来看，如果说这么多年以来我的建筑设计作品还算有些特点的话，很大程度上是因为我既学习了西方的理论思想，又能够跳开西方的条条框框，从自己的文化中寻找到一些设计的支点。这其中，中国式的思维方式、中国传统哲学以及一些中国传统的文艺创作理论都使我受益良多。我常感慨，中国文化中值得我们细细体会，可以转化为设计理论的内容太多了。但是这些内容，或者还并不为人所知，或者还只是停留在某些建筑师的个人感悟。比起体系完整的西方建筑理论，我们对于中国建筑理论的思考深度还远远不够，已有的研究还非常不系统。刘先觉先生写过一本《现代建筑理论》[1]，是一本让人开卷有益的好书。我放在案头，时时翻阅。但遗憾的是，这么厚一本书却没有一点中国人的东西。我们中国建筑师、建筑学者真的不能对当代建筑理论的发展作出一点贡献吗？

所以我认为，当下对中国建筑的研究不能只是满足于表态性的言论，或者枝枝节节的观点，而是要立根，从深度上进行挖掘，要重视对哲学与美学的探究，厘清建筑理论的思想脉络。除此之外，我还认为应该重视对建筑创作机制的研究。我自己将建筑创作理论视为沟通建筑学中一些宏观问题与设计中具体问题的中介性理论，这也是我多年来重点思考的一个领域。我希望自己的研究能突破对建筑设计的一些常规认识，做一些不一样的探索。

费：关于中国建筑设计要走自己的路。可能很多人都会觉得这个提法不陌生，因为中国近现代建筑发展历程中一直不乏这方面的讨论。但从近现代建筑的发展历程来看，历史上很多次这样的讨论最后都不了了之了，中国建筑的发展道路最后还是会回到西方主义的老路上去。您对这个问题是怎么看的？今天再谈这样的话题，又有什么不同吗？建立自己的理论体系真能解决这个问题吗？

程：上面已经谈过了，自20世纪初以来，中国建筑的发展兜兜转转，总也摆脱不了西方主义的影响。那么为什么会这样呢？我认为，最根本的需要从西方和我们自己这两方面来分析。

就西方来说，它们是以一种比较平滑的方式进入现代社会的。现代文化与西方传统之间有着千丝万缕的联系。这使得西方建筑的发展能建立在一个连绵不绝的文化基础之上，而呈现出很强的体系性与生命力。例如前几年备受瞩目的非线性建筑，看似横空出世，但是其体现出来的复杂性和非理性思维，不正是对20世纪后半期当代哲学与美学思想的最好呈现吗？所以西方建筑能席卷整个世界，固然与全球化有关，但也是有其内在文化的原因。

反观我们自己，我们是在积贫积弱、坚船利炮的双重压力下被迫进入现代社会的，付出的隐性代价就是与自己文化的割裂。而没有了自己的文化与价值取向作为依傍，建筑也好，其他文化创作领域也好，要走出一条自己的道路几乎是不可能的。这也就是为什么中国建筑发展之路会如此曲折的原因。

不管从上面哪个方面来看，问题的症结都落在了文化上，所以关键还是要建立自己的文化体系。但需要特别指出的是，这个"文化"不应该是传统文化，而是基于当下现实的中国当代文化。唯有这样的中国文化体系，才能对今天的文艺创作起到指引与支撑作用。建筑也好，其他艺术创作领域也好，才能在自身文化土壤的滋养下，不断地生发出一些新的方向，变得更为丰富而有生命力。

费：您刚才提到中国文化的认识，并特别指出中国文化并不等同于中国"传统"文化，我觉得这是一个特别重要的观点，能不能请您就这一点再深入地谈一下。

程：这两年学习传统已经成为一股社会热潮，我们经常在各种媒体上看

4. 加纳国家大剧院
Ghana National Theatre, Accra

5. 马里会议大厦
Conference Building Mali, Bamako

6. 浙江美术馆
Zhejiang Art Gallery

到满是"弘扬传统文化"这样的话语。对于这样的话语我内心深处是不赞同的。今时不同于往日，社会在发展，人们生产与生活方式都在不断改变。仅仅靠弘扬传统文化，就能解决当代问题了吗？

传统是什么？传统是我们的文化之根，也是流淌在我们血液中的文化基因。我们之所以重视传统文化，目的并不是要复古，而是希望延续自己的文化基因，然后转换创新，向前发展。但是如果我们简单地将传统与现代中国文化混为一谈，那就是泛化了传统，也抹杀了当代的价值，更否定了中国文化不断转换更新的可能，实际上就是一种文化上的不自觉。在很多时候，传统已经成了一个挡箭牌，遮盖了问题的核心。这个核心就是我们是一个有着丰厚文化传统的国家，但是我们更需要建构一个符合我们当下情境的中国当代文化。

现在有一个倾向，就是不分主次、不分好坏地把传统文化中表面的、往往也是最浅显的东西拿出来混充中国文化，或者将继承传统等同于肤浅的符号拼贴。最突出的表现是直接把坡屋顶、马头墙，以及汉服祭孔、读《弟子规》当作"传统"来大力"弘扬"。前一阵，我在央视国际频道看到一个节目，将昆曲与嘻哈音乐结合在一起，作为中国当代文化介绍给全世界。我很喜欢昆曲，也很关心昆曲在当代的发展。但看到这样一种生硬的拼贴与混搭，我感到特别尴尬与别扭。可是这样一种对待传统的态度，现在可真不少见。这既是对我们自己文化传统的严重贬低，也是对文化转换创新的简单化理解。

对于传统，我赞成冯友兰先生提出的"抽象继承"，即透过那些物质与非物质的遗存去理解传统的内在精神、价值判断与认知模式等，将其中仍有生命力的东西融入今天的价值、思想体系中。中国文化如一条奔腾的大河，它是从传统中来的，但必须融合现代的内容，必须对当代问题作出明晰的回应。我想说，这个千呼万唤的"中国当代文化"，是不会从天上掉下来的，也不是从故纸堆里扒拉出来的。它需要各个领域，也包括建筑设计领域，从自身的实践出发，深入比较反思，转化创新，逐步形成新的、有自己特色的当代中国文化，当然这是一个长期的、动态发展的过程。

费：您特别强调文化的独特性，但是另一方面我们也看到所谓"独特性"都是相对的，没有一种文化是可以完全不受外来文化影响的。就拿中国文化来说，它其实是一个特别开放包容的体系，中国文化的历史演进是建立在对他者不断学习吸收的基础上，逐渐形成的，您对此是怎么看的呢？

程：你的观点完全正确，我一直也是这么想的。文化与文明都是流动的，是彼此影响的。我从不反对向西方学习。相反，我认为中国建筑的发展离不开跨文化对话，我们应该坚持走跨文化发展的道路。但是在面对传统与现代、东方与西方的时候，关键还是要对我们自己在世界文化发展中所处的位置有一个"自觉"。

我曾经说过，人类文化的发展是一个由时间和空间、传统和现代构成的多维坐标体系[2]。这其中既存在一个在既有文化基础上自我更新的过程，也有在外来文化的影响下裂变发展的阶段。就中国文化来说，2000年来，尽管不断有外来文化的影响，历史上中国文化基本上是沿纵向——通过自我更新来实现发展的。宋元以后一直到"五四"，中国文化已经无法通过自身的更新发展而衰落了，于是才有了西方文化的涌入。中国文化在这样的一个过程中，通过横向交流，逐步实现了走向现代中国文化的突破，这是一个客观事实。可以说在不同时段，每种文化都会面临一个偏于纵向还是偏于横向发展的选择。

现在距离"五四"已经过去近百年了，中西文化已经出现了新的走向和趋势。甚至像一些持西方中心主义的学者如亨廷顿也说，西方文明正从它的高峰滑落，进入一个迷惘、焦灼的时期。在这样一个时期，不是说西方就不会出现优秀的文化作品了，但是他们思想生产的活跃度与深度确实在下降。很多问题在他们自己内部的体系内很难找到答案。如果说过去百年主要是我们学习西方，那么现在更多的是相互的学习。"跨文化"已经成为当代一个非常重要的关键词。对于我们自己来说，更应该抓住当下的历史机遇，自觉地调整我们在十字坐标上的运动轨迹，在跨文化的交流碰撞中实现中国文化的转换更新。所以我说要建立中国自己的建筑理论体系，并不是要拒绝西方的影响，而是希望大家能客观地看待中西方文化发展的历程与它们之间的关系，既不要狂悖自大，也不要妄自菲薄。我认为在文化自觉的基础上，抱着平等对话的态度，通过跨文化研究，中国建筑的发展空间无限。

费：您说的"跨文化发展"，我相信很多人都会赞同。但是真正要做到跨文化并不容易。在"学习他者"之后，还要"成为自己"就更难了。您觉得应该如何做到这一点呢？

程：这确实不容易，怕就怕学习他者不到位，自己又没有东西，那么就只能永远跟在人家身后人云亦云了。所以我很赞同乐黛云先生提出的观点，当下"中国文化更新的希望就在于深入理解西方思想的来龙去脉，并在此基础上重新理解自己"[3]。只有做到这一点，才能谈得上"跨文化"发展。在我看来，这句话中的"深入"与"重新"这两点非常关键。

所谓"深入"，就是要对西方现代建筑应该做历史的、全面的观察，而不应为一个时期、一种流派所局限。就西方当代建筑来说，自20世纪初至今，它是不断演变的，这其中既有片面狂悖，也有不断调整的自我补偿，有益的经验往往存在于那些观点完全相反的流派之中。因此，我们不仅要研究形形色色的西方建筑思潮的兴衰得失，还要关注它的发展走向，更要以一种批判的态度去深入地分析这些理论，这对于建构我们自己的理论体系十分重要。所谓"重新"，就是在学习西方的同时，在当代语境下，以一种选矿淘金的态度去辨识传统，努力把自己的东西搞清楚，"重新"认识自己。有的时候，我们是"只缘身在此山中"，恰恰缺乏对自己的认识。所以这两年，我的博士研究生的研究基本上是两条线索都有：既有研究西方建筑理论的，如建构、现象学、陌生化理论；也有研究中国建筑理论的，比如对江南建筑文化的研究，对于"通感""直觉"这种中国式创作思维的研究。

在跨文化对话中，我们可能会发现，中西方所论述的很多问题都有关系。双方的观点常常是一体两面、相反相成的。这个相反相成很重要，不是绝对对立，而是互补互生。当我们从相反相成的角度来观察中、西方的思想理论线索的时候，我们就会生成一种更为全面、客观的认识。其中最关键的一个词就是"化"：融通中外，转换提升，化入心中。我觉得如果能做到这一点，那就能够"成为自己"。

费：每个专业领域对中国文化都有着不同的理解和阐释，您极力想在建筑中呈现的"中国文化精神"又是什么呢？

程：这是一个必须回答的问题，但回答这个问题，还是要在中西文化比较的基础上来理解。我一直有一个观点：做建筑创作，必须要有思想，要有哲学和美学支撑，否则就站不住脚，容易被人同化。也许有人会问，一些知名的西方建筑师并不是个个都有理论啊？对，他们也许不是个个都有理论，但是他们的工作与西方哲学、美学思想线索之间的关系是清晰可辨的。他们

7. 建川博物馆（俘虏馆）
Prisoner of War Museum (POW) of Jianchuan Museum Complex

8. 李叔同（弘一大师）纪念馆
Master Hongyi Memorial Hall

9. 海宁博物馆
Haining Museum

的作品不是毫无由来、横空出世的，西方的哲学与美学思想从古典时期开始，一直到后来的现代主义、后现代主义，一直绵延不绝地为他们的建筑创作提供了思想与文化的滋养。相比之下，近代以来中国文化的断裂，对中国的文艺创作领域的影响是巨大的。它导致了中国建筑处于一种彷徨无所依靠的尴尬境地。

因此我自己的建筑创作理论希望是从文化的角度入手，探讨中国当代建筑的哲学、美学基础，从而解决建筑设计的问题。尽管这些思考很多涉及思想与文化层面，但是其最初发端却是来自于我自己的建筑实践历程，是一点点慢慢积累，逐步形成一个相对完整的构架。

经过对自己实践与思考历程的反复思考与梳理，我提出了以"境界"为哲学基础、以"意境"为美学特征、以"语言"为媒介的建筑创作理论。我不想说它是对中国文化精神最好的诠释，我只想说它对我的创作非常重要。这些理解使我早就摆脱了"传统"的困扰，也走出了"西方"的桎梏，找到了自我，获得了创作的自由，我很满意自己这样一种创作状态。

当然，我在不同场合反复强调一点：这只是我的"一家之言"。其实，每一个建筑师都可以根据自己的文化背景和性格素养特点，从不同角度去理解中国文化，去建构自己的认知体系。多元的认识与不断求变的创作实践相结合，中国现代建筑就能茁壮成长为一片枝叶丰满、多姿多彩的美丽森林。

关于中国建筑理论建构的探索

费：您提的这个以"境界"为哲学基础、以"意境"为美学特征、以"语言"为媒介的建筑理论，似乎与西方的语言哲学是相对的，是这样吗？

程：你可以说它们是相对的。20世纪西方哲学的语言学转向，其潮流席卷全球，对很多学科都产生了影响。在建筑学领域，也是各种"语言"的天下，耳熟能详的像符号学语言、类型学语言、模式语言、空间句法，以及近几年兴起的参数化语言等，影响不容忽视。

从我个人的创作经历来说，我却对这几十年来主要从语言的角度去认识世界、认识建筑的观点持怀疑态度。比如建筑学中的模式语言与类型学，从理论上来说都没有问题，但放到现实中去，却常常碰壁。为什么呢？我认为最主要的原因就在于这一类理论对于建筑设计的理解过于抽象化与简单化了。建筑设计不是靠几十或者几百种类型、模式就可以囊括的。用这样的方式去解读、分析建筑与日常生活，就无法获得对这个世界的整体认识，也就和这些理论的初衷背道而驰了。

相比西方对"语言"的重视，中国人的"大美不言""天何言哉"，表达了我们对于语言、对于这个世界的不同认识。为什么"不言"？那是因为一旦言说，理性高涨，感性消退，整体性的世界也就分崩离析。但是我们又绝不是不讲理性，而是对那种条分缕析的简单理性持非常审慎的态度，我们总是希望能通过复杂理性的转化提升，回归高度的整体性之中。

所以我一直在思考能不能跳出语言，从中国哲学出发，找到另一种对世界、对建筑的认知方式，这就是由"境界""意境""语言"这3个层次构成的建筑理论。以"境界"为哲学本体，就是要从自然、自我的角度出发，追求主客体的和谐共生；以"意境"为美学特征，是要从生命的意志与情趣出发，超越物象的束缚，追求时空中的情感共鸣；以"语言"为工具与手段，是要在认识到语言规范性作用的同时，也能摆脱程式化或者非理性的倾向，寻找其与精神性追求的耦合。这样一种建筑理论，算是我从自己的创作经历出发，提出的一家之言吧。

费：也就是说，您所提出的这个建筑理论是建立在中国哲学基础上的，这是其与西方建筑理论的最根本区别。那么从哲学层面来看，"境界"的内涵是什么呢？

程：西方哲学一直是从逻辑、分析的角度来认知世界。希腊哲学家在探索万物起源的时候，就是将世界分解为最简单的单元，并通过对一个个局部的研究，达到认识世界的目的。毕达哥拉斯的"万物皆数"，柏拉图的"理式"，体现的都是这样的一种观点。影响到当代建筑学，出现类型学与模式语言也是非常自然的事情。

相比西方对解析的重视，在中国古代先哲的眼里，世界是一个混沌如一、密不可分的整体。《道德经》中说"有物混成，先天地生""道生万物""万物归于道"等，论述的就是这个意思。在这样一种整体论思想的基础上，我们会自觉地将世界视为是一个相互联结、密不可分的结构整体。面对这样复杂的结构整体，我们在认知方法和时间活动上也与西方截然不同。我们更注重从整体上去加以把握，并对它的内在关系作出综合性的描述、诠释与安排，使得这个复杂结构的各部分达到一种自然而然、恰到好处的状态。正如有学者在解析王国维《人间词话》的时候说，"妙手造文，能使其纷沓之情思，为极自然之表现"[4]，这就是我所说的境界。

这样一种境界首先是从人的感性实践中逐渐积累沉淀出来的，反映了人与自然之间的整体联系，是人与自然之间关系的内化，体现了"浑然天成""天人合一"的状态。这是一种"最高智慧"，也是我在建筑创作中所追求的一个层次、一种状态。

事实上，这样一种哲学认知不仅对建筑创作，而且对所有文化领域的探索都有着重要意义。当下，社会文化发展日趋复杂，科技也以一种超出人们意想的速度迅猛发展。但是人与自然、万物之间的这种本质性关系不会改变。我认为，这样一种对世界、包括对建筑的哲学认知，可以帮助我们摆脱当代社会中物我之间、人与自然之间的紧张关系，实现世界的和谐永续发展。

费：您这样的解释使我对"境界"有了一个基本了解。您刚才说的"整体性"，似乎是"境界"说的一个重要方面？

程：是这样的，"境界"所体现的首先就是一种整体性的思想观念。反映到建筑领域，就会生发出一种整体综合、自然有机的建筑观，并在此基础上形成一种整体性建筑设计思维模式，而这对建筑创作有着至关重要的影响。

与之相反的是一种西方式的单向思维模式，比如过去我们说"功能决定形式"，后来又出现了"形式包容功能"，再到近年来对于结构、材料因素的强调等。在我看来，这些思维模式实际上将建筑创作的机制理解得过于简单化了。建筑创作是一个复杂的过程，要考虑的问题很多，除了卡彭在《建筑理论》[5]中说的形式、功能、意义，以及结构、文脉、意志之外，还需要关注建筑所处的物质与文化环境，考虑建筑的经济、安全、生态、节能，以及市场分析、业态策划等因素，有些项目还要考虑对城市与区域发展的影响。这些复杂因素之间的相互交织，不是简单的 A 决定 B 这样的单向逻辑可以概括的，也不可能将这些因素预设为"基本范畴"与"派生范畴"[5]。所以在创作过程中，我更愿意将这些复杂因素之间的关系，看成是由一个个相对独立节点所构成的多维立体网络。建筑设计就是思想不断在这个网络中游走的过程。在此过程中，通过一个恰到好处的切入点，激活整个网络，使得各

10. 杭州铁路新客站外景
New Railway Station of Hangzhou

11. 杭州国际假日酒店
Holiday Inn Hangzhou

12. 南京博物院（二期工程）
Nanjing Museum (Phase II)

个问题都能得到相对合理的解决，就是一种整体性的思维方式。所以古人说，"文章本天成，妙手偶得之"。"浑然天成"就是我们在建筑创作中需要这样一种整体性的思维模式，它显然与西方式强调逻辑分析不同，是一种关照整体的复杂理性。

以我自己早年设计黄龙饭店的经历为例，在设计之初，我就意识到这个项目的复杂性，需要综合地考虑好功能、流线、结构、经济、管理等一系列问题，特别是建筑与自然环境之间的关系，文化心理成了解决这个复杂问题网络的切入点。最后的设计通过没有先例的单元成组分散式布局，不仅取得了建筑与自然之间、现代功能与文化心理之间的一种微妙平衡，而且也创造性地解决了酒店设计中一系列复杂问题，包括流线长度、客房布局、公共空间设计、酒店管理等。这一次经历给了我很大的信心。我一直在说当时这个方案之所以能在与境外单位竞标中胜出，主要原因不是我的设计水平有多高，而是要归功于中国传统哲学中的整体性思维给予的启发。

费：您说的这个"境界"，似乎具有一定模糊性与不确定性？

程："境界"的价值就在于它在确认世界是一个有机整体的同时，注意到了它的模糊性与不确定性。《道德经》中说"道之为物，惟恍惟惚。惚兮恍兮，其中有象；恍兮惚兮，其中有物……"。在恍兮惚兮中去探寻物象，在"玄而又玄"中去寻找"众妙之门"，这是中国哲学追问世界本原的一种特殊思维方式。

我认为这样一种认识对于文艺创作有很大的积极意义。因为这个世界在很多方面并不是非黑即白、非对即错的。有时候对于精确性、逻辑性的片面追求，会束缚我们的思维，反而会阻碍我们对事物的认识与创造。这一点在建筑创作中，表现得十分明显。

建筑创作是一个在混沌一体的世界中去体悟、去辨识、去创造的过程。这个过程在最初肯定是模糊、不确定的，答案也不可能只有一个。随着思考的深入，各种要素、想法在不断复合、碰撞之后，逐渐相互交织、混沌一片的状态中，显露出内在的秩序。而这样一种过程与状态不仅适用于建筑设计，在其他艺术创作领域，甚至科学研究领域都有着类似的特点。爱因斯坦就说过，"人所能体验的最美和最深刻的东西就是神秘，它是一切艺术、科学中所有深刻追求的基础"[6]。

现在有一种将建筑设计科学化的倾向，对此我不敢苟同。建筑设计当然应该重视科学研究，这里面必然包括对一些客观性要素的定性与定量分析。但仅靠科学分析，或者简单的量化、加权、排序，是不可能生成一个优秀设计作品的。在我看来，设计者对所有主、客观因素的一个整体、综合理解远比一些量化分析更为重要。我希望"境界"说的模糊性、不确定性，能帮助大家摆脱束缚与教条，给建筑设计领域的发展带来更为广阔的空间。

费：那么建筑创作中的这种模糊性与不确定性，是否也会影响到我们对建筑设计学科的理解呢？

程：我认为是肯定的。建筑设计学科虽然有科学性的一方面，但它不是纯科学。它所具有的模糊性与不确定性，就是区别于自然科学与其他工程技术学科的一个重要方面。建筑设计学科的发展，必须建立在自身学科特点的基础上。不能将适用于理工科的评价与考核方式，比如说 SCI 论文数量，套用到建筑设计学科。但现在这样的现象却非常普遍，这其中的原因很多。我想最根本的一条可能就在于，建筑设计学科被当下的科研评价体系所裹挟。有的时候为了突出建筑学科在整个学科体系中的地位，或者为了考核或排名，片面强调建筑设计学科的科学性，而丢失了自身的学科特点。如果建筑设计学科的主要目标是培养建筑师，那么我想建筑设计学科的发展还是应该围绕建筑创作的特点，否则不但建筑设计学科的未来发展让人担忧，对建筑创作的繁荣也是不利的。

费：您的这些观点，与当前主流的建筑设计思想似乎有所不同。我想这样一种观点应该不仅仅局限于建筑创作领域，也会对当下的建筑设计理论与建筑设计学科的发展产生影响，您对此是怎么看的呢？

程：在我看来，建筑学是一门随着时代前进、不断动态发展的学科。正如我在 20 年前所提出的："社会发展速度愈来愈快，摆在我们面前的问题愈来愈多：纳米材料、虚拟空间会给建筑带来怎样的影响？宽带网、数字化对人们生活方式、行为方式所产生的影响怎样反馈给建筑？跨文化发展带来的碰撞交流，又将使人们的审美方式和价值取向产生什么样的变化？[7]"可以说，建筑学的内涵和外延已经发生，也正在不断发生种种变化。正因为此，"任何一种流派都只能从一个时期、一个侧面去认识建筑，以一隅之见拟万端之变是不可能的。在建筑创作领域，没有金科玉律，任何一种流派、理论都只能是一家之言"[8]。建筑学领域的许多理论思想都与其所处的时代背景，包括科学技术水平、对世界的认识等有关。建筑师与建筑学者对这些理论思想应该持审慎的态度，应该关注时代与文化发展与这些理论思想的关系。从当代的角度来看，一些西方经典理论，很多可能已经不适用了。未来建筑设计学科将如何发展？这些都需要我们放开胸怀，共同探索。

所以我谈"境界"不是要提一个很"玄"的概念，而是针对当下建筑学发展中的一些问题，希望对当下建筑学领域中一些比较僵化的认识能有所反思与触动。将千差万别的观念"定于一尊"，或者将某些过去"经典"奉为圭臬都是不足取的。

中国建筑设计领域的发展，必须破除盲从与迷信，特别是破除对西方理论的迷信，要敢于建构一家之言，这对当下中国建筑设计理论与创作领域的发展有着重要意义。

费：能不能请您从自己的角度出发，谈一谈您对"境界"与建筑创作的关系的感受？

程：在《庄子·达生篇》里曾经讲过"梓庆为鐻"的一个故事。说的是一个名叫"庆"的木匠非常善于用木做鐻，见过他作品的人无不惊叹其鬼斧神工。有一次鲁侯问他："子何术以为焉？"他回答说我也没有什么特别的技巧，但是每次我要准备做鐻的时候，就先要斋戒几天，"齐以静心"，借此忘却一切外界烦扰及自身技巧，直至"忘吾有四肢形体"。然后"入山林"，观察各种木料的天性，直到鐻的形象已经呈现于我的眼前之后才开始动手，如是一气呵成。

在这个故事里，工匠对鐻的认识不是从解析开始的，而是将自己与其所要创作的对象，放到整个天地、自然中去一同进行体悟琢磨，所以才有在他的创作里也没有"自然""自我""形式""材料"的区分，所有这些始终都是浑然一体的。"器之所以疑神者"就在于"以天合天"。

我认为这段话所描述的就是建筑创作的一种非常好的状态。在这里，建筑创作的思考过程不是西方式的，在逻辑、方法加持下去寻求答案，而是像"梓庆为鐻"那样自然而然、水到渠成。在这个过程中需要设计者将个人全部的技巧、经验、知识、感受都融入进去，以"我"之自然去合"物"之自然。而当人的自然与物的自然相契合，达到"浑然天成"的状态，就可以称

13. 上海公安局办公指挥大楼
Commanding Center of Shanghai Public Security Bureau

14. 宁波高教园区图书信息中心
Ningbo Library and Information Center, University Zone of Ningbo (NLIC)

15. 绍兴鲁迅纪念馆
Shaoxing Lu Xun Museum

为是一种境界。

以前张在元先生曾以"泰宁尺度"来评价我的建筑创作，大家可能认为他说的是建筑尺度的问题，其实不然。张先生说的"尺度"指的是我对建筑设计过程中整体性与恰到好处的分寸感的把握。有了这样的"尺度"，才能谈得上"境界"。我自己常感慨，陆机在《文赋》中所说的"精骛八极，心游万仞……观古今于须臾，抚四海于一瞬"，就是对建筑创作进入境界之后，那种酣畅淋漓之感的最好描述。

费："境界"这个词容易与"意境"混在一起，在您的理论体系里面，"境界"是本体，"意境"是美学基础，它们两者的关系与区别是什么呢？

程：王国维的《人间词话》中是存在"境界"与"意境"混用的现象，但我认为这两者是不同的。"境界"是哲学层面的思考，而"意境"侧重于美学层面，谈的是从中国文化角度对什么是"美"的一种理解。从哲学层面来说，西方是一种唯理体系，目的在于探析世界的基本结构、秩序，因此西方人对美的追求往往从逻辑、结构、数理这些角度去理解。而中国哲学则是一种生命体系，旨在突破外在的束缚，了解与体验生命的意趣。因此我们对于什么是美的认识，也与西方有着本质不同。所以说"境界"与"意境"之间有关系，但并不是同一个事儿。

中国人对美的理解是建立在生命体验的基础上，对美和艺术进行经验性的感悟，并通过整体的概念进行意会性的表述。因此我认为中国人对美的理解具有主体性、整体性、意会性。所谓主体性，是说中国人对美的认识，总是以自我为主体，"美"并不具有独立的意义，它是为情感和内容服务的。整体性指的是，中国人总将美作为一个浑然的整体进行研究，不像西方美学那样对美和艺术进行解剖式的分析。所谓"意会性"指的是中国人用直观、感性进行体验、体会，而不善于用明确的概念进行表述，也不像西方人那样对美的标准进行定量分析。

这3个特征决定了中国人对什么是美的看法，很少受到具体形象的束缚，也很少对形式去进行明确的规范。谢赫的画论六法中的"气韵生动、骨法用笔、应物象形、随类赋彩、经营位置、传移模写"，谈的都不是具体的形式原则。我们总是希望突破具体的、在时间与空间上的有限性的"象"而达到"境"。因此谢赫会说，"若拘以体物，则未见精粹，若取之象外，方厌膏腴"①。

刘禹锡会说，"境生于象外"②。意境是对具体形象的突破，在时间与空间上都趋向于无限的"象"，是中国古代艺术家常说的"象外之象""景外之景"。在当代社会中，这样一种对美的理解，或许可以帮助我们摆脱消费文化对于感官的奴役，重新回归到人内在的精神需求。

费：那么您认为建筑中的"意境"之美指的是什么呢？

程：这个问题必须放到具体例子中才好谈。以江南园林来论，它的意境在哪里呢？我认为有几点是很重要的。其一，文人造园其实是营造一个小世界，这个小世界由山、水、树、建筑（亭台楼阁）所组成。它融入了一年四季，也融入了天、地、人的关系，"一花一世界，一木一浮生"。园林意境的真谛在于超越那些片段式的景观，而映射一个完整的精神世界。这个世界是具体的，也是抽象的，也是主体与客体高度统一的。我们在园林中感受到了自然与人的高度融合。其二，园林的空间往往是有限的，但是造园者却擅长用有限的空间映射出无限的世界。所谓"轩楹高爽,窗户虚邻,纳千顷之汪洋,收四时之烂缦"，这种突破具体形式束缚的"意境"，方为园林的真趣。从园林来看，我们不妨说，所谓意境可能就在于突破物象、形式的束缚，以有限来映射无垠。

以我自己设计的建川战俘馆为例，在建筑创作中，形式被有意识地弱化了，我想突出的是一种氛围与意境——压抑、扭曲、悲怆……这是我对战俘这一特殊人群心理的表达，并以此来打动观众，强化建筑的艺术感染力。

应该说这种对美的理解，在一些西方当代艺术家与建筑师的作品中，同样也有所体现。例如在意大利画家乔治·莫兰迪的油画作品中，通过那些反复排列、形状单纯的瓶子，我也能感受到一种超越物象而达到某种意境的类似追求。在现代主义建筑大师巴拉干的建筑作品中，在那些静谧的几何空间中，我们也能感受到这种"意境"之美。只是在西方语境中，对于这样一种美学体验是缺乏自觉的。在西方，意境从来不是一个独立的美学范畴，这也是在英语中找不到与"意境"相对应词语的原因。但是在中国人的美学认知中，对于"意境"的追求却一直是艺术创作的应有之义。我希望当代的中国人能够重新学会欣赏这种"意境"之美，也希望中国建筑师能够更自觉、更充分地去表达意境之美，这也可以说是中国建筑师的天然优势。

费：通过您前面的解释，我对"境界""意境"有了了解。在您的建筑理论中，境界、意境、语言是共同构成一个体系的。您之前讲过，语言是载体，而非本体。这与西方哲学有着很大差异。能不能请您就这一点，针对语言再展开谈一下？

程：在西方，20世纪是语言哲学的天下。海德格"语言是存在之家"，德里达说"文本之外无他物"。这几年数字语言的盛行，更确立了语言哲学在西方的统领地位。反映到建筑学中，现在也是各种语言的天下，从几十年前的模式语言到现在的参数化，影响甚广。

对此我是这么看的。"语言"包含语意,特别是它对"只可意会不可言传"的创作机制做了理性分析，这一点是值得肯定的。但同时我们也应该看到，它忽视了万事万物之间的深层联系，忽视了人们的文化心理和情感，很难完整地解释和反映建筑创作的实际。针对这些方面，所以我才提出境界、意境、语言这样一个三位一体的理论体系。

我认为，建筑创作如果沿着西方以语言为本体的思路发展，极易走入偏重"外象"的形式主义歧路。从20世纪后半叶开始，以语言为本体的哲学认知与后工业社会文明相结合，西方文化出现了从追求"本原"，逐渐转而追求"图像化""奇观化"的倾向。在建筑设计领域中，出现了一些以追求建筑形式的感官刺激为目的的趋势。面对这一现象，我们是否也应该思考一下，这种以语言为哲学本体，注重外在形式的思想是否有它的局限性。

所以我认为，是时候反思一下西方总是在"语言"与"形式"上兜圈子的思路了。与西方人对于将形式作为哲学本体特别重视相比，我们发现中国人很少孤立地谈语言与形式问题，不管是"以形写神"，还是"言以表意""言以寄理"，都是将语言、形式和它要传达的情感、意蕴结合在一起。这说明了中国人对于"形""意"（意境）"理"（境界）这三者之间关系的理解。那么这样一种观点是不是对当代中国哲学、美学建构以及中国当代建筑理论的发展会有所启发呢？

费：能不能请您具体谈一谈，在建筑创作中，您是怎么思考"语言""意境""境界"这三者关系的？

程：首先，正如前面已经提过的，语言不能脱离其所表达的意境、境界单独存在。建筑创作中推敲语言，追求形式，最终目的还是在于营造意境、表达境界。一旦将语言剥离开考虑，它也就丧失了创作性与活力，或者成为

16. 绍兴市民广场
Shaoxing People's Square

17. 浙江宾馆商务别墅
Business Villa of Zhejiang Hotel

18. 联合国国际小水电中心
International Center on Small Hydro Power

一种纯粹以刺激人感官为目的的"奇观性"建筑，或者如一度流行的"国际式"建筑，陷入僵化与重复之中。

其二，境界、意境、语言这三者之间并不存在主次、高低之分，也不是一个谁决定谁的关系。一方面确实是"术以载道""言以表意"，但是另一方面也是"道术相长"，它们之间是互为一体的。语言与形式创新没有了哲学与美学的支撑固然不行，但是如果缺少了语言的承载，建筑就成为一种抽象的观念，少了生活与情感的热度。

其三，既然形式是载体，那么在创作中，建筑师为了更好地表达自己的理念，可选择的手段应该是多种多样的。特别是较之"意""理"层面的相对稳定，"语言"会随着时代的发展而不断变化，建筑师需要在充分掌握古今中外建筑语言的基础上，不断转化创新。所以我对在建筑设计中用元素拼贴的方式，或者简单地以某种风格，比如说现在流行的新中式，来表达中国文化精神的做法是非常不赞同的。每个项目的条件不同，又怎么能用千篇一律的语言来表达呢。

前一阵，很多人都在议论"奇奇怪怪"的建筑。我认为要全面理解总书记对"奇奇怪怪"建筑的批评，不能以此来否定建筑创新。建筑创新就要"守正出奇"，关键是要"奇"得有"情"有"理"。以我刚刚完成的温岭博物馆为例，这个建筑用了非线性的形式语言塑造成一块山石的形态，有人可能看着会觉得有些"奇怪"。但正是因为这样一种形式语言的选择与推敲，才使得这个博物馆能于周边混乱市井之中而自成一种气场，同时也与当地的石文化取得呼应。所以如果我们把这个建筑放到它所在的城市环境与文化背景中去看，我们就一点都不会觉得它有什么"奇怪"。同时也因为这个建筑形式背后的文化底蕴，才使得它与一些程式化的非线性造型区别开来，具有了一种中国韵味。

就像我前面提过的"境界""意境""语言"这三者共同构成的建筑创作理论，是我自己在中国传统哲学、美学的基础上，从自己的实践与思考出发，面向复杂现实问题，提出的一家之言。虽然未必成熟，但是提出这样的一家之言进行讨论，不仅对当下现实，而且建筑学科，以至于文化科技领域的发展都是有益的。中国建筑理论研究迫切需要摆脱西方"经典"的桎梏，在平等进行跨文化对话的基础上，从我们的文化背景出发，从自身的实践出发，去寻找答案。而且类似的讨论与探索应该多一些，更大胆一些。唯此，中国建筑设计才能够真正取得生机与自由。

费：一般来说，实践建筑师关注理论研究的并不多，更不要说形成这样系统、完整的理论思想了。所以最后我想请您谈谈，作为一个实践建筑师，您是如何一步步地建立您的理论思考架构的呢？

程：我很多对建筑的思考都是来自于自己的建筑实践历程。它并不是一开始就有一个构架，而是日积月累，逐渐形成了一个相对完整的认识。这其中有几个工程项目对我个人思想的成长是有很大影响的。

以20世纪80年代初设计的黄龙饭店为例，当时我们设计团队之所以能在国际竞标中胜出，中国传统哲学中的整体性思维给予我很大的启发。通过这个工程项目，我开始认真地思考，如何将中国传统的哲学与美学思想转化到建筑设计中去。这期间，东西方文化之间的关系，是我一直在思考的问题。从20世纪80年代中叶的"立足此时、立足此地、立足自己"[9]，到后来的《在历史与未来之间的思考》[2]，以及10年以后的《在历史与未来之间的再思考》[10]，我对自己的建筑创作之路应该怎么走，已经有了一个比较清晰的认识。

这期间我遇到了加纳国家大剧院、杭州铁路新客站、弘一法师纪念馆、建川战俘馆、浙江美术馆等不同类型的建筑设计项目。在创作过程中，我深刻地体会到了，一个好的建筑方案需要综合地处理好功能、形式、场地、技术、经济、文化等各种要素之间的关系。这里面是一个理性与非理性的不断复合、相互转化的过程，而建筑设计就是要找到那个恰到好处的平衡点。这引起了我对建筑创作机制的思考，深化了我对中国文化精神的认识。在2000年初，我从认识论、方法论与美学理想这3个层面出发，提出了"天人合一""理象合一""情景合一"这样一个比较完整的、属于中观层次的建筑创作理论。到了2010年以后，我开始思考能不能从哲学美学层面出发，打通古今、融合东西，建立一种基于中国当代情境的建筑理论体系，这就是境界、意境、语言3个层面。所以我的理论不是事先有个构架、体系，而是从工程总结中去思考方向问题、哲学美学问题、方法问题。在思想渐渐萌发、形成的同时，到书本中去寻找一种印证与共鸣，然后交织形成自己的思想框架。

我一直认为，建筑设计领域中的"思"不是抽象的玄思，它应该是与"做"

结合在一起，是实践中的思考，也是思考中的实践。因此理论与实践之间的关系，不是简单的理论指导实践，而是"知行合一"。我希望我所提出的这些观点、理论，能引起建筑学者、建筑师的讨论，也希望有更多人能关心这些研究，投身到这些研究中去。建筑这门学科具有模糊性和不确定性，没有标准答案，没有金科玉律，所以探索的空间很大，魅力也就在这个地方。对于建筑学的未来，包括中国建筑的未来，我还是很乐观的。

注释

① 见谢赫《古画品录》。
② 见刘禹锡《董氏武陵集记》。

参考文献

[1] 刘先觉. 现代建筑理论 [M]. 北京：中国建筑工业出版社, 2001.
[2] 程泰宁. 在历史和未来之间的思考 [J]. 建筑学报, 1989(2): 39-41.
[3] 乐黛云, 阿兰·李比雄. 跨文化对话 4[M]. 上海：上海文艺出版社, 2000.
[4] 许文雨. 钟嵘诗品讲疏 人间词话讲疏 [M]. 成都：成都古籍出版社, 1983.
[5] (英) 戴维·史密斯·卡彭. 建筑理论 [M]. 北京：中国建筑工业出版社, 2007.
[6] (美) 阿尔伯特·爱因斯坦. 信仰自白 [M]// 爱因斯坦文集. 北京：商务印书馆, 2010.
[7] 程泰宁. 程泰宁作品选：1997-2000[M]. 北京：中国建筑工业出版社, 2001.
[8] 程泰宁. 我的建筑哲学 [M]// 中国当代建筑师（第一卷）. 天津：天津科学出版社, 1988.
[9] 程泰宁. 立足此时、立足此地、立足自己 [J]. 建筑学报, 1986(04): 11-14.
[10] 程泰宁. 面向未来，走自己的路——在历史和未来之间的再思考 [J]. 建筑学报, 1997(1): 7-10.

LANGUAGE, ARTISTIC CONCEPTION AND INTELLECTUAL STATE
AN INTERView WITH CHENG TAINING ON HIS ARCHITECTURAL THOUGHTS

Fei Yishan: Fei   Cheng Taining: Cheng

originally published in *Architectural Journal*, issue 10, 2018

At the beginning of 2017, I began to assist Academician Cheng Taining in the preparation of the book *Language and Intellectual State*. Initially, I had planned to establish a general understanding of his architectural thoughts through one or two Q&A sessions. Unexpectedly, such interviews were conducted dozens of times afterwards. Each interview was like a garden with divergent paths, and many ideas were different from the mainstream of architecture field, which often made me feel confused and bewildered at the beginning. As the discussion progressed, I gradually came to understand these threads of thought and their relationship with contemporary architecture, and gained some insights of my own.

As Academician Cheng himself said, "The key point in architecture research is 'transformation', to integrate the Chinese and the foreign, to convert and upgrade, to transform into the heart". For Cheng, all his thinking is based on the soil of practice, which is the mutual reference and advancement of practice and thinking, so it has a different value and significance from the general theoretical research. During the discussion, Cheng reminded many times that architecture is characterized by complexity, multiplicity and openness, which is difficult to be sorted out by simple rationality in a compartmentalized manner. Therefore, the best presentation of the network of Cheng's thoughts is perhaps his own architectural works and those fleeting words themselves. I hope that the expediency for the sake of this article does not detract from the intellectual value of the words themselves.

This paper is a selection and collation of dozens of discussions in the past two years. The first half of this article is mainly a reflection on the current development direction of architectural design. The second half is a presentation of the architectural thought system of "Language, Artistic Conception and Intellectual State" proposed by Academician Cheng Taining.

Reflections on the development direction of Chinese architecture

Fei: You are regarded as a practice architect who has completed many impressive architectural works. But in recent years, you seem to have suddenly become concerned with theoretical research, writing several books and speaking out more in academic fields. What is the reason behind this shift?

Cheng: As an architect, I have always kept the habit of thinking while creating. In the past, whenever I completed an important architectural work or gained insights in my creative practice, I would also write some theoretical articles and send them to the Architecture Journal for publication. For me, practice and theory always complement each other, and there is no sudden change in between. However, in the past two years, I did pay more attention to theoretical research than I did in the past. This is mainly because the contradictions in architectural design and related fields are particularly prominent nowadays, and theoretical research is particularly important to solve these problems. It is also because after so many years of accumulation, I have a clearer understanding of many problems in architectural creation, and it is time for me to talk about them in depth.

Fei: What exactly do you mean by this "contradiction"?

Cheng: I believe that one of the most fundamental contradictions plaguing the current field of Chinese architectural design remains the issue of values. What is good architecture? We have lost a clear judgment.

The alienation of architecture is one of the most important aspects. A good building should meet the basic requirements of applicability, safety, ecology and energy saving, as well as technical and economic rationality, which is what the *Beijing Charter* of the International Union of Architects (UIA) calls "back to the basics". However, at present, many buildings are deviating from the social functions and cultural values they are supposed to carry, and are being alienated into a propaganda tool for some people to boast and show off their wealth, a commodity to be consumed and entertained. Under the influence of such a trend, it is difficult to evaluate architecture by normal value standards, and to a large extent, it leads to confusion in architectural design thinking.

Another important aspect is the cultural dwarfing of oneself in the face of the west. I always believe that the influence of western architecture on the development of Chinese architecture, both positive and negative, should not be underestimated. In the past two years, there are more people talking about cultural self-confidence, but the idea of "still the foreign moon and fuller" is still deeply rooted both in the society and in our own field. Many of our leaders and developers are still seeking strange things from foreign countries. In the design bidding of large projects, Chinese architects still need to be "close" to foreigners. Recently, a cultural project of a mega-city in the south was tendered, and only foreign architects were allowed to participate. In the field of architectural design, it is also a common phenomenon in the past 30 years to take "western" as "modern" and to copy and imitate as "convergence". This phenomenon of talking about cultural self-confidence on the one hand, but being held hostage by the West on the other, is it not a prominent reflection of the contradictions of the present time?

I believe that without solving these contradictions, it is impossible to achieve the overall improvement of China's architectural design level.

Fei: I can understand your concern about the phenomenon of blindly admiring foreigners, but at the same time I have some doubts. For example, you have repeatedly talked about the difference in values between the east and the west. What is the impact of this on the future development of Chinese architecture? Is it possible to divide architectural design into Chinese and western?

Cheng: People may have a misunderstanding that I talk about "Chinese architecture" only out of a sense of national pride, and that I deliberately want to strike a discordant note with western architecture. In fact, this is not the case at all. My own architectural education started from the western Beaux-Arts system. After that, I went through the renewal of various theoretical trends and methodological systems, such as modernism, postmodernism, tectonic, and parametric, together with everyone else. For a long time when I was young, I was obsessed with western architecture, and in the 1980s I wrote articles calling for greater communication between China and the west. It would be impossible to say that I was not influenced by these western architectural theories and methods. But it is precisely because of my love and concern that I have come to know both their goodness and their limitations.

I believe that we must take a historical and comprehensive view of western architecture, and not be limited by one period or one school. At present, the development of Western architecture is facing many problems and contradictions. A while ago, the famous Japanese architect Maki Fumihiko said something in an interview to the effect that the modernism of the 1950s was a big ship that everyone wanted to ride on, but now the big ship is gone and there are only countless pieces floating on the sea. As a member of the same generation, I feel the same way. After the west entered the post-industrial society, under the impact of post-modern culture, western architecture has shown a tendency of fragmentation and anti-rational development. Under such circumstances, it would be undesirable for us to take some fragment floating in the sea as our direction. Therefore, when I talk about the value construction of Chinese architecture, I am not intentionally trying to sing against western architecture, but to explore our own direction, which is the only way to get rid of rootless floating and blindly following the trend.

I would like to say that both Chinese and western cultures, including architectural culture, are at a certain point of development. Diverse exploration and cultural reconstruction are the characteristics of this era, and we should have a full understanding of this.

Fei: As for the future development of Chinese architecture, many architects and architecture scholars may now have different views, for example, they may be more concerned with the exploration within their own fields and less concerned with what belongs to "China" and what belongs to "the west". What do you think about this?

Cheng: In my opinion, the difference between China and the west is certainly a difficult knot that haunts many Chinese architects. The collision between the two sides in politics, economy and culture is objective. How to understand the relationship between China and the west is a problem that many Chinese architects face almost every day. And the in-depth consideration of this issue constitutes an important part of Chinese architectural theory.

So why are some people "uninterested" in exploring this difference between Chinese and Western cultures? I think there are two possible reasons for this.

One is because Chinese architects have long been under the influence of westernism. Since the modern era, every Chinese's knowledge system and way of thinking have been branded with deep traces of westernism. In the case of architecture, from the first day we studied architecture, we were taught a set of architectural education from the west. It is because we have been in it for a long time that we tend to "treat other countries as our homeland".

In addition, another more important reason may be that this issue is too big and there are many objective factors, which are difficult to say clearly. So nowadays, there are many architecture scholars and architects who deliberately avoid these topics. It is not that they do not feel these problems, but they may feel that this topic is too big and not solved by architecture and architectural design. Many problems are not clear even after talking about them, so they are not willing to talk about them.

For the former point, architects, especially young architects, need to think independently. In time, they will naturally understand what the difference between Chinese and western architectural cultures is. For the latter view, although I can understand it, I cannot accept it. Since the problem already exists objectively, we must face it seriously. We cannot deliberately avoid discussing this issue just because it is big and difficult. It is precisely because it is big and difficult that we need to work on it little by little from our own field of knowledge, and accumulate small victories into big ones, so that we can eventually make a breakthrough. We often say that the world should develop in the direction of diversity. We should think about what Chinese architects and architecture scholars can contribute to this diversified world. This is not too much to ask, right?

Fei: What kind of role can your recent emphasis on theoretical research play in the development of Chinese architecture today? Or how do you see the current relationship between theoretical research and architectural practice in China?

Cheng: For Chinese architecture today, the role of theoretical research is indeed very important. You can think back. It's not that Chinese architects haven't wanted to make something of their own in the last hundred years. For example, in the 1930s, the "inherent Chinese forms" were advocated; in the 1950s, the "national style" was advocated; in the 1980s, the "recapture of the ancient capital style" was also a hot topic in the architectural circles at that time; in recent years, there is no lack of discussions on such topics as "new and Chinese" and "new Chinese style". However, due to the lack of a strong theoretical system to support, people only used formal language to counter formal language and populism to reject foreign cultures. In the end, they were always only lively for a while before being quickly submerged under the influence of westernization. This shows that without theoretical support, relying only on national pride and "blindly following the trend", it is impossible to achieve the real development of Chinese architecture. In my opinion, practical innovation, theoretical reflection and value system construction are complementary to each other. Many of the problems reflected at the practical level are related to the evaluation system, which in the end must also rely on theoretical construction to solve. Without theoretical support, cultural confidence becomes a slogan, and practical innovation cannot be truly realized.

Chinese architecture needs to construct something conceptually and ontologically, to form the ideological and theoretical core of Chinese architecture, and to gradually establish its own architectural theory system. The world architecture is like a colorful forest, and Chinese architecture is also one of the "subjects". We cannot just be satisfied with grafting a small branch on the big tree of western architecture. Whether it is the coconut tree in Hainan or the poplar in the northwest, they all grow independently from their own land. In the past few years, I have participated in some theoretical research projects, chaired the completion of the Chinese Academy of Engineering project "the Present and Future of Architectural Design in Contemporary China", accepted some interviews and wrote some articles. In addition to the hope that I can explore the way for the development direction of Chinese architecture, I also hope that through these works, I can draw people's attention to these issues and let more people devote themselves to the work in this field.

Fei: There are still a lot of Chinese architecture scholars and architects who are paying real attention to China's "own" problems, what do you think about these researches?

Cheng: In recent years, a number of scholars and architects have done solid work in some research on specific issues, such as the application of materials and the inheritance and development of traditional craftsmanship. A number of quite distinctive architectural works have also emerged. In the study of architectural theory, including the practice of architectural creation, cutting from some specific issues is an approach that is easier to understand and better to start with. My own exploration of the theory of architectural creation also started from specific problems. These studies on specific problems can be regarded as the accumulation before discussing some larger problems. Without these studies, our discussion of Chinese architectural theory would have no foothold.

However, we also need to see that the current research on Chinese architecture lacks systematization and depth. Many researchers automatically regard "Chinese" architectural theory as a "complement" to current architecture. "Complement" means that it can be deep or shallow, and does not pursue systematization. Therefore, many theoretical explorations are always superficial and do not seek for a system. It is this view that has greatly limited the development of Chinese architectural theory.

From my personal creative experience, if my architectural design works over the years have some characteristics, it is largely because I have studied western theoretical ideas, but I have also been able to look beyond western rules and regulations to find some design fulcrums from my own culture. I have benefited a lot from the Chinese way of thinking, traditional Chinese philosophy, and some traditional Chinese literary and artistic creation theories. I often feel that there are so many elements of Chinese culture that are worthy to be experienced in detail and can be translated into design theory. However, these elements are either not well known or remain in the personal perceptions of some architects. Compared with the complete system of western architectural theory, our thinking about Chinese architectural theory is still far from deep enough, and the existing research is still very unsystematic. Liu Xianjue has written a book *"Modern Theories of Architecture"*[1], which is a good book that is useful for people to read. I put it on my desk and read it from time to time. But unfortunately, there is nothing Chinese in such a thick book. Is it true that we Chinese architects and architecture scholars cannot make any contribution to the development of contemporary architectural theory?

Therefore, I believe that the current research on Chinese architecture should not just be satisfied with epigrammatic remarks or branching views, but should establish roots, dig from depth, pay attention to the exploration of philosophy and aesthetics, and clarify the ideological vein of architectural theory. In addition, I also think we should pay attention to the study of the mechanism of architectural creation. I myself regard the theory of architectural creation as a mediating theory that bridges some macro issues in architecture and specific issues in design. This is also an area that I have focused on for many years. I hope my research can break through some conventional understanding of architectural design and do some different exploration.

Fei: Regarding Chinese architectural design to follow its own path, many people may find this reference not unfamiliar, as there has been no shortage of discussions in this regard in the course of China's modern architectural development. But judging from the development history of modern architecture, many times in history such discussions were finally shelved, and the development path of Chinese architecture would eventually go back to the old path of westernism. What do you think about this issue? Is there any difference when we talk about this topic today? Can we really solve this problem by establishing our own theoretical system?

Cheng: As already discussed above, since the beginning of the last century, Chinese architecture has gone round and round in circles, unable to escape the influence of westernism. So why is this so? In my opinion, the most fundamental thing to analyze is from both the west and ourselves.

In the case of the west, they entered modern society in a relatively smooth way. There is an inextricable link between modern culture and western tradition. This has enabled the development of western architecture to be based on an unbroken cultural foundation that is highly systematic and vital. For example, the non-linear architecture that has attracted much attention in the past few years seems to have emerged out of nowhere, but isn't its embodied complexity and irrational thinking the best presentation of contemporary philosophical and aesthetic thought in the second half of the 20th century? The fact that western architecture has swept the world is certainly related to globalization, but it also has its own internal cultural reasons.

On the contrary, we were forced to enter modern society under the double pressure of poverty and weakness, and the hidden price we paid was the severance from our own culture. Without our own culture and values, it is almost impossible for architecture and other cultural creation fields to follow a path of their own. This is the reason why the road of Chinese architecture development is so twisted.

Regardless of the above aspects, the crux of the problem lies in culture, so the key is to establish our own cultural system. However, it should be noted that this "culture" should not be the traditional culture, but the contemporary Chinese culture based on the current reality. Only such a Chinese cultural system can guide and support the creation of literature and art today. Architecture and other fields of artistic creation can be nourished by their own cultural soil and constantly give birth to new directions, becoming richer and more vital.

Fei: You have just mentioned the understanding of Chinese culture and pointed out in particular that Chinese culture is not the same as Chinese "traditional" culture. I think this is a particularly important point, and I would like to ask you to talk about it in more depth.

Cheng: In recent years, learning about tradition has become a social craze, and we often see words like "promoting traditional culture" in various media. Deep down, I don't agree with such words. Today is different from the past, society is developing, people's production and lifestyle are changing. Can we solve the problems of our time just by promoting traditional culture?

What is tradition? Tradition is the root of our culture and the cultural gene that flows in our blood. The reason why we value traditional culture is not to restore the past, but to continue our cultural genes and then transform and innovate to move forward. But if we simply confuse tradition with modern Chinese culture, we are generalizing tradition, erasing contemporary values, and denying the possibility of continuous transformation and renewal of Chinese culture, which is actually a kind of cultural unconsciousness. In many cases, tradition has become a shield that obscures the core of the problem. This core is that we are a country with a rich cultural tradition, but we need to construct a contemporary Chinese culture that fits our current context.

Nowadays, there is a tendency to take the superficial and often the most superficial aspects of traditional culture and mix them up with Chinese culture, or to equate inherited traditions with superficial symbolic collages, regardless of priority and quality. The most prominent manifestation of this is the direct "promotion" of sloping roofs and horse-head walls, as well as the offering of Confucius in Chinese costume and the reading of the *Disciple's Rule* as "traditions". A while ago I saw a program on CCTV International that combined Kun Opera with hip-hop music and introduced it to the world as contemporary Chinese culture. I like Kun Opera very much and I am very concerned about its contemporary development. But I felt particularly embarrassed to see such a raw collage and mash-up. But such an approach to tradition is not uncommon nowadays. This is both a serious devaluation of our own cultural traditions and a simplistic understanding of cultural transformation and innovation.

I agree with Feng Youlan's idea of "abstract inheritance", that is, to understand the inner spirit, value judgment and cognitive mode of the tradition through the material and immaterial remains, and to integrate what is still alive in it into today's value and thought system. Chinese culture, like a great rushing river, comes from tradition, but it must be integrated with modern content and must provide a clear response to contemporary issues. I would like to say that this "Chinese contemporary culture", which is called for by all, will not fall from the sky, nor will it be pulled out of a pile of paper. It requires all fields, including the field of architectural design, to start from their own practice, compare and reflect deeply, and transform and innovate. Only in this way can a new contemporary Chinese culture with its own characteristics be formed gradually. Of course, this is a long-term, dynamic development process.

Fei: You emphasize the uniqueness of culture, but on the other hand we can see that "uniqueness" is relative. There is no culture that is completely free from foreign influence. In the case of Chinese culture, it is in fact a particularly open and tolerant system. The historical evolution of Chinese culture is based on the continuous learning and absorption of others. What do you think about this?

Cheng: Your point is exactly right, and I have always thought so. Cultures and civilizations are fluid and influence each other. I have never been against learning from the West. On the contrary, I think the development of Chinese architecture cannot be separated from cross-cultural dialogue, and we should stick to the path of cross-cultural development. But when facing tradition and modernity, east and west, the key is to have a "self-awareness" of our own position in the world cultural development.

As I have said before, the development of human culture is a multidimensional coordinate system consisting of time and space, tradition and modernity[2]. There is both a process of self-renewal based on existing cultures and a phase of fission development under the influence of foreign cultures. In the case of Chinese culture, for two millennia, despite the constant influence of foreign cultures, Chinese culture has basically developed along the vertical axis -- through self-renewal. After the Song and Yuan dynasties and up to the May Fourth Movement, Chinese culture was no longer able to develop through its own renewal and decline, and only then did the influx of western culture occur. It is an objective fact that Chinese culture has gradually achieved a breakthrough towards modern Chinese culture through horizontal exchanges in such a process. It can be said that at different times, each culture faced a choice of vertical or horizontal development.

Now that nearly a century has passed since the May Fourth Movement, a new direction and trend has emerged in Chinese and western culture. Even some western-centric scholars, such as S. P. Huntington, have said that western civilization is slipping from its peak into a period of confusion and anxiety. In such a period, it is not that there will be no outstanding cultural works in the west, but the activity and depth of their intellectual production is indeed declining. Many questions are difficult to find answers to within their own internal systems. If we have been learning from the west for the past hundred years, we are now learning more from each other. "Interculturalism" has become a very important keyword in our time. For ourselves, we should seize the current historical opportunity, consciously adjust our trajectory on the cross-coordinates, and realize the transformation and renewal of Chinese culture in the cross-cultural exchange and collision. Therefore, when I say that we should establish China's own architectural theory system, I do not mean to reject the influence of the west, but I hope that we can objectively look at the history of the development of Chinese and western cultures and the relationship between them, and not to be either arrogant or presumptuous. I believe that on the basis of cultural self-awareness and with an attitude of equal dialogue, there is unlimited room for the development of Chinese architecture through cross-cultural research.

Fei: I believe many people would agree with you when you say "cross-cultural development". But it is not easy to be cross-cultural. It is even more difficult to "learn from the other" and then to "become yourself". How do you think this should be done?

Cheng: This is not easy indeed. The fear is that if we do not learn from others, and we do not have anything, then we will have to follow others forever. Therefore, I agree with Yue Daiyun's view that "the hope for the renewal of Chinese culture lies in a deeper understanding of the origins of western thought and a new understanding of ourselves on this basis"[3]. Only when this is done can we talk about "cross-cultural" development. In my opinion, the two points of "deep" and "new" in this statement are very crucial.

By "in-depth", we mean that we should take a historical and comprehensive view of western modern architecture, and not be limited by a period or a school. As far as western contemporary architecture is concerned, it has been evolving since the beginning of the last century. It has been a mixture of one-sided paradox and self-compensation with constant adjustment. Useful experiences are often found in schools of thought that have completely opposing views. Therefore, we should not only study the rise and fall of various western architectural trends, but also pay attention to their development, and analyze these theories in depth with a critical attitude, which is very important for the construction of our own theoretical system. The so-called "new" means that while studying the west, in the contemporary context, we should identify the tradition with a kind of gold-seeking attitude, try to clarify our own things, and "reacquaint" ourselves. Sometimes, we are "lost in the heart of the very place", and we lack the knowledge of ourselves. Therefore, in the past few years, the research of my doctoral students has basically been on both fronts: there are those who study western architectural theories, such as tectonic, phenomenology, and theories of strangeness; there are also those who study Chinese architectural theories, such as the study of Jiangnan architectural culture, and the study of the "synaesthesia (Tonggan)" and "intuition (Zhijue)", which are Chinese style of creative thinking.

In cross-cultural dialogue, we may find that many of the issues addressed by China and the west are related. The views of both sides are often two sides of the same coin and opposite to each other. This opposition is important, not in the sense of absolute opposition, but in the sense of complementarity and mutuality. When we look at the intellectual and theoretical threads of China and the west from the perspective of the opposite, we will develop a more comprehensive and objective understanding. One of the most crucial words is "transformation": integrating Chinese and western, transforming and upgrading, and into the heart. I think if we can do this, then we can "become ourselves".

Fei: Each professional field has a different understanding and interpretation of Chinese culture. What is the "spirit of Chinese culture" that you are trying to present in your architecture?

Cheng: This is a question that must be answered, but answering this question is still to be understood on the basis of the comparison between Chinese and western cultures. I always have a view that to do architectural creation, one must have ideas, philosophical and aesthetic support, otherwise it will be untenable and easy to be assimilated. One may ask, some famous Western architects do not all have theories, right? Yes, they may not all have theories, but the relationship between their work and the threads of western philosophical and aesthetic thought is clearly discernible. Their works did not come out of nowhere. Western philosophical and aesthetic thought from the classical period to the later modernism and post-modernism has been providing the intellectual and cultural nourishment for their architectural creations. In contrast, the rupture in Chinese culture since modern times has had a tremendous impact on the field of Chinese literary and artistic creation. It has led to an awkward situation in which Chinese architecture is at a loss and has nothing to rely on.

Therefore, my own theory of architectural creation hopes to explore the philosophical and aesthetic foundations of contemporary Chinese architecture from a cultural perspective, so as to solve the problems of architectural design. Although many of these reflections involve ideological and cultural aspects, their initial beginnings come from my own architectural practice, which has been gradually accumulated and gradually formed a relatively complete framework.

After repeatedly thinking and sorting out my practice and thinking process, I proposed such a theory of architectural creation with "intellectual state" as the philosophical foundation, "artistic conception" as the aesthetic feature, and "language" as the medium. I do not want to say that it is the best interpretation of the spirit of Chinese culture. I just want to say that it is very important to my creation. These understandings have enabled me to get rid of the trouble of "tradition" and to get out of the shackles of "west". I have found myself and gained the freedom to create. I am very satisfied with such a state of my creation.

Of course, I have repeatedly stressed on different occasions that this is only my "one school of thought". In fact, every architect can understand Chinese culture from different perspectives and construct his own cognitive system according to his own cultural background and character. With the combination of diversified understanding and ever-changing creative practice, Chinese modern architecture can grow into a beautiful forest with rich and colorful branches.

Exploration on the construction of Chinese architecture theory

Fei: This theory of architecture that you have mentioned, with "intellectual state" as the philosophical basis, "artistic conception" as the aesthetic feature, and "language" as the medium, seems to be opposed to the western linguistic philosophy, is it so?

Cheng: You could say they are relative. The linguistic turn in Western philosophy in the 20th century has swept the world and has had an impact on many disciplines. In the field of architecture, too, various "languages" are at work. Familiar ones such as semiotic language, typological language, pattern language, spatial syntax, and the recent emergence of parametric language have had an impact that cannot be ignored.

From my personal creative experience, I am however skeptical of the viewpoint of understanding the world and architecture mainly from the perspective of language in these decades. For example, the pattern language and typology in architecture are fine from the theoretical point of view. However, when put into reality, they often hit a wall. Why? I think the main reason is that this kind of theory is too abstract and simplistic for the understanding of architectural design. Architectural design cannot be encompassed by dozens or hundreds of types and patterns. If we interpret and analyze architecture and daily life in such a way, we will not be able to gain a holistic understanding of the world, which is contrary to the original purpose of these theories.

Compared to the Western emphasis on "language", the Chinese phrase "the great beauty does not speak" and "does Heaven speak" expresses our different understanding of language and the world. Why "not speak"? It is because once we speak, reason rises, sensibility recedes, and the world as a whole fall apart. However, we are by no means irrational, but we take a very cautious attitude toward the simple rationality of compartmentalization. We always hope to return to a high degree of wholeness through the transformation and elevation of a complex rationality.

Therefore, I have been thinking about whether we can go beyond language and find another way of perceiving the world and architecture from Chinese philosophy. This is the theory of architecture which consists of three levels: "intellectual state", "artistic conception" and "language". Taking "intellectual state" as the philosophical ontology, it is to pursue the harmonious coexistence of subject and object from the perspective of nature and self. With "artistic conception" as the aesthetic feature, it is to pursue the emotional resonance in time and space from the will and interest of life, beyond the bondage of objects. Using "language" as a tool, we want to recognize the normative role of language, but at the same time, we also want to get rid of the tendency of typology or irrationality, and find its resonance with the pursuit of spirituality. Such a theory of architecture is my own opinion from my own creative experience.

Fei: In other words, this theory of architecture that you propose is based on Chinese philosophy. This is the most fundamental difference between it and the western architecture theory. What is the connotation of "intellectual state" from the philosophical point of view?

Cheng: Western philosophy has always perceived the world from a logical, analytical point of view. When the Greek philosophers explored the origin of everything, they broke the world into its simplest units and understood the world by studying one partial part after another. From Pythagoras' "everything is number" to Plato's "eidos", they all embody such a view. It is also very natural to influence contemporary architecture with the emergence of typology and pattern language.

In contrast to the western emphasis on analysis, in the eyes of ancient Chinese philosophers, the world is a chaotic and inseparable whole. This is what the *Tao Te Ching* means when it says, "There is a thing formed in chaos existing before Heaven and Earth," "The Tao begets all things," and "All things return to the Tao". On the basis of such a holistic mindset, we will consciously regard the world as an interconnected and inseparable structural whole. Faced with such a complex structural whole, our cognitive approach and temporal activities are also very different from those in the west. We pay more attention to grasp it as a whole and make comprehensive descriptions, interpretations and arrangements of its inner relations, so that the parts of this complex structure can reach a natural and appropriate state. As some scholars said in their analysis of Wang Guowei's *Jen-Chien TZ'u-hua*, "the masterful hand of the writer can make his varied feelings and thoughts become extremely natural expressions"[4]. This is what I mean by the intellectual state.

Such a state is gradually accumulated and precipitated from human sensual practice, reflecting the overall connection between man and nature. It is the internalization of the relationship between man and nature, reflecting a state of "natural harmony" and "unity of heaven and man". This is the "highest wisdom", a level and state that I pursue in my architectural creation.

In fact, such a philosophical perception is important not only for architectural creation, but also for the exploration of all cultural fields. Nowadays, social and cultural development is becoming more and more complex, and technology is developing rapidly at a speed beyond people's expectation. However, the essential relationship between human beings and natural beings will not change. I believe that such a philosophical perception of the world, including architecture, can help us get rid of the tension between matter and self, and between people and nature in contemporary society, while achieving a harmonious and sustainable development of the world.

Fei: Your explanation has given me a basic understanding of "intellectual state". The "wholeness" you mentioned just now seems to be an important aspect of "intellectual state", is that so?

Cheng: What "intellectual state" embodies is first of all a holistic ideology. Reflected in the field of architecture, it will give rise to a holistic and organic view of architecture, and on this basis to form a holistic architectural design thinking mode, which has a crucial impact on the creation of architecture.

The opposite is a western-style one-way thinking model, for example, we used to say that "function determines form", and then "form accommodates function", and in recent years, the emphasis on structural and material factors. In my opinion, these modes of thinking actually over-simplify the mechanism of architectural creation. In addition to the form, function, meaning, structure, culture, and will mentioned by David Smith Capon in his *Architectural Theory* [5], it is also necessary to pay attention to the physical and cultural environment in which the building is located, consider the economic, safety, ecological, and energy-saving factors of the building, as well as market analysis and business planning, and in some projects, consider the impact on urban and regional development. The intertwining of these complex factors cannot be summarized by the simple one-way logic of A deciding B. It is also impossible to pre-determine these factors as "basic categories" and "derived categories"[5]. Therefore, in the process of creation, I prefer to see the relationship between these complex factors as a multi-dimensional network composed of relatively independent nodes. Architectural design is the process of constantly moving ideas through this network. In this process, through an appropriate entry point, the whole network is activated, so that each problem can be solved in a relatively reasonable way, which is a holistic way of thinking. That's why the ancients said, "The essay was originally made naturally without artificiality, and was obtained by chance by a person of great skill". The "natural" is the holistic mode of thinking that we need in the creation of architecture. It clearly differs from the western-style emphasis on logical analysis and is a complex rationality that cares for the whole.

Taking my own experience in designing the Dragon Hotel as an example. I realized the complexity of this project and the need to consider a series of issues such as function, flow, structure, economy, and management in a comprehensive manner at the beginning of the design process. In particular, the relationship between the building and the natural environment, as well as the cultural psychology, became the entry point to solve this complex network of problems. The final design not only achieved a delicate balance between architecture and nature, modern function and cultural psychology through the unprecedented decentralized layout of units in groups, but also creatively solved a series of complex problems in hotel design, including flow length, room layout, public space design, hotel management, etc. This experience gave me a lot of confidence. I kept saying that the main reason why this proposal won the competition with overseas bids at that time was not because of my high level of design, but the inspiration given by the holistic thinking in traditional Chinese philosophy.

Fei: This "state" you mentioned seems to be somehow ambiguous and uncertain, is that right?

Cheng: The value of "intellectual state" lies in its recognition of the world as an organic whole while noting its ambiguity and uncertainty. In the *Tao Te Ching*, it is said that "Who can of Tao the nature tells? Our sight it flies, our touch as well. Eluding sight, eluding touch, the forms of things all in it crouch; Eluding touch, eluding sight, there are their semblances, all right……" In the trance to explore things, in the "mysterious of the mysterious" to find the "door of all wonders". This is a special way of thinking for Chinese philosophy to ask about the origin of the world.

I think such an understanding has great positive significance for the creation of literature and art. Because the world is not black or white, right or wrong in many aspects. Sometimes the one-sided pursuit of precision and scientific logic can fetter our thinking and hinder our understanding and creation of things. This is evident in the creation of architecture.

Architectural creation is a process of perceiving, discerning and creating in a world of chaos and unity. This process is certainly ambiguous and uncertain at the beginning, and there cannot be only one answer. As we think deeper, various elements and ideas will gradually reveal the inner order in the intertwined and chaotic state after continuously compounding and colliding. Such a process and state are not only applicable to architectural design, but also has similar characteristics in other fields of art creation and even scientific research. Einstein said, "The most beautiful and profound thing that man can experience is mystery, which is the basis of all profound pursuits in all arts and sciences"[6].

Nowadays, there is a tendency to scientize architectural design, which I disagree with. Architectural design should of course pay attention to scientific research, which necessarily includes the qualitative and quantitative analysis of some objective elements. However, it is impossible to generate an excellent design work by scientific analysis alone, or by simple quantification, weighting and ranking. In my opinion, a holistic and comprehensive understanding of all subjective and objective factors is far more important than some quantitative analysis. I hope that the vagueness and uncertainty of the "state" can help people get rid of the constraints and dogmas, and bring a broader space for the development of architectural design.

Fei: So, does this ambiguity and uncertainty in architectural creation also affect our understanding of the discipline of architectural design?

Cheng: I think so. Although the discipline of architectural design has a scientific aspect, it is not a pure science. The ambiguity and uncertainty of it, is an important aspect that distinguishes it from natural science and other engineering technology disciplines. The development of architectural design discipline must be built on the basis of its own disciplinary characteristics. The evaluation and assessment methods applicable to science and engineering disciplines, such as the number of SCI (Science Citation Index) papers, cannot be applied to architectural design disciplines. However, such a phenomenon is very common nowadays, and there are many reasons for this. I think the most fundamental one may lie in the fact that the architectural design discipline is held hostage by the current research evaluation system. Sometimes, in order to highlight the position of architecture in the whole discipline system, or for the purpose of assessment or ranking, the scientific nature of architectural design discipline is unilaterally emphasized, and its own disciplinary characteristics are lost. If the main goal of architectural design discipline is to cultivate architects, then I think the development of architectural design discipline should still focus on the characteristics of architectural creation, otherwise not only the future development of architectural design discipline is worrying, it is also unfavorable to the prosperity of architectural creation.

Fei: These views of yours seem to be different from the current mainstream architectural design thinking. I think such a view should not only be limited to the field of architectural creation, but also have an impact on the development of architectural design theory and architectural design discipline nowadays. What do you think about this?

Cheng: In my opinion, architecture is a discipline that evolves dynamically with the times. As I said 20 years ago, "The speed of social development is increasing, and the questions before us are becoming more and more numerous: How will nano-materials and virtual space affect architecture? How will the impact of broadband network and digitalization on people's lifestyles and behaviors feedback to architecture? What kind of aesthetic and value changes will be brought about by the cross-cultural development?"[7]. It can be said that the connotation and extension of architecture have been and are being changed. Because of this, "any school can only understand architecture from one period and one side, and it is impossible to compare all the changes with a single view. In the field of architectural creation, there is no golden rule, and any kind of school or theory can only be the opinion of one."[8] Many theoretical ideas in the field of architecture are related to the background of their time, including the level of science and technology and the understanding of the world. Architects and architecture scholars should take a prudent attitude toward these theoretical ideas and should pay attention to the relationship between the times and cultural development and these theoretical ideas. From a contemporary perspective, many of the classical western theories may no longer be applicable. How will the discipline of architectural design develop in the future? All these need to be explored together with an open mind.

So, I talk about "intellectual state" not to mention a very "metaphysical" concept, but to address some of the current problems in the development of architecture, hoping to reflect on and touch some of the more rigid understanding in the field of architecture today. It is not desirable to "settle on a single authority" a wide range of different concepts or to take certain "classics" of the past as a guideline.

The development of China's architectural design field must break away from blind obedience and superstition, especially from the superstition of western theories, and dare to construct a family of opinions, which is of great significance to the development of China's architectural design theory and creation field nowadays.

Fei: Can you tell us how you feel about the relationship between "intellectual state" and architectural creation from your own perspective?

Cheng: In *Chuang Tzu Ta Shang,* there is a story about a carpenter named Khing who makes bell-stand. The story is about a carpenter named "Khing" who was very good at making bell-stand out of wood, and everyone who saw his work was amazed at his skillful work. Once the marquis of Lu asked him, "by what art you had succeeded in producing it?" He replied that he doesn't have any special skills, but whenever he wants to make a bell-stand, he first fasts for a few days, "to compose my mind", so as to forget all external disturbances and him own skills, until he " had forgotten all about myself; — my four limbs and my whole person". Then he "went into the forest " and observe the nature of the various woods until the image of the bell-stand was already in front of his eyes, and then he would begin to do it.

In this story, the artisan's understanding of instr. does not begin with analysis, but with his own perception of the whole world and nature together with the object he wants to create, so that there is no distinction between "nature," "self," "form," and "material" in his creation, all of which are always integrated. The reason why "[the] spirit was thus engaged in the production of the bell-stand" lies in the fact that "[the] Heaven-given faculty and the Heaven-given qualities of the wood were concentrated on it " .

I think what this passage describes is a very good state of architectural creation. Here, the thinking process of architectural creation is not the western way, seeking answers with the support of logic and methods, but a natural and watery process like "Khing makes bell-stand". In this process, the designer needs to incorporate all his personal skills, experience, knowledge and feelings, and to combine the nature of "matter" with the nature of "self". And when the nature of human and the nature of things fit together and reach the state of "natural", then it can be called an intellectual state.

In the past, Zhang Zaiyuan used the "Taining scale" to evaluate my architectural creation, and people might think he was talking about the scale of architecture, but in fact, he was not. What Zhang meant by "scale" is my grasp of the integrity and the right sense of proportion in the architectural design process. With such "scale", we can talk about "intellectual state". I often feel that the words of Lu Ji in his *Wen Fu,* " The spirit flies beyond the faraway land, and the mind travels to a height of ten thousand feet. […] View the past and the present in a moment, caress the four seas in a moment", is the best description of the feeling of pleasure after the architectural creation enters the state.

Fei: The word "intellectual state" is easily confused with "artistic conception". In your theoretical system, "intellectual" is the ontology and "artistic conception" is the basis of aesthetics. What is the relationship and difference between them?

Cheng: In Wang Guowei's *Jen-Chien TZ'u-hua*, there is a confusion between "intellectual state" and "artistic conception", but I think they are different. "Intellectual state" is a philosophical reflection, while "artistic conception" focuses on the aesthetic level, talking about an understanding of what "beauty" is from the perspective of Chinese culture. On the philosophical level, western philosophy is a rational system, which aims at analyzing the basic structure and order of the world, so westerners tend to understand the pursuit of beauty from the perspective of logic, structure and mathematics. Chinese philosophy, on the other hand, is a system of life, aiming to break through external constraints to understand and experience the meaning of life. Therefore, our understanding of beauty is fundamentally different from that of the West. So, there is a relationship between "intellectual state" and "artistic conception", but it is not the same thing.

The Chinese understanding of beauty is based on life experience. It is an empirical perception of beauty and art, and an intentional expression through the concept of wholeness. Therefore, I believe that the Chinese understanding of beauty has subjectivity, wholeness, and meaningfulness. By subjectivity, I mean that the Chinese understanding of beauty is always self-centered. "Beauty" does not have an independent meaning, but serves the emotion and content. Holistic means that the Chinese always study beauty as a whole, unlike western aesthetics, which analyzes beauty and art in an anatomical way. The term "sensemaking" refers to the fact that the Chinese experience and appreciate beauty intuitively and sensually, but are not good at expressing it in clear concepts, nor do they analyze it quantitatively as westerners do.

These three characteristics determine the Chinese view of what is beautiful, which is rarely bound by specific images or clear norms on form. Xie He's Six points to consider when judging a painting, which include "Spirit Resonance, Bone Method, Correspondence to the Object, Suitability to Type, Division and Planning, and Transmission by Copying", are not about specific formal principles. We always want to break through the specific, limited "image" in time and space and reach the "realm". That is why Xie He would say: "If we confine ourselves to the physical, we will not see the essence, if we take it outside the image, we will be tired of the richness"[①]; Liu Yuxi would say "aesthetic conception transcends concrete objects described"[②]. The realm is a breakthrough of concrete image, which tends to be infinite in time and space, and is what ancient Chinese artists used to call "the image beyond an image" and "the scene beyond a scene". In contemporary society, such an understanding of beauty may help us get rid of the slavery of consumer culture to the senses and return to the inner spiritual needs of people.

Fei: What do you think the beauty of "artistic conception" in architecture refers to?

Cheng: This question must be put into concrete examples before we can talk about it. In the case of the Jiangnan Garden, where is its artistic conception? I think there are a few points that are important. First, the literati built the garden to create a small world, which consists of mountains, water, trees, and buildings (pavilions). It is integrated into the four seasons of the year, and also into the relationship between heaven, earth and people, "to see a world in a flower, to feel a life in a wood". The true meaning of garden artistic conception is to go beyond those fragmentary landscapes and map a complete spiritual world. This world is both concrete and abstract, and is also highly unified between subject and object. We feel a high degree of integration of nature and people in the garden. Second, the space of the garden is often limited, but the garden maker is good at mapping an infinite world with a limited space. As the saying goes, "The porticos in front of the hall are tall and straight, and the windows face the vast space, enough to receive a thousand hectares of ocean, the rosy time of the seasons", this conception that breaks the boundaries of specific forms is the true interest of the garden. From the garden point of view, we might say that the so-called conception may lie in the breakthrough of the constraints of the objects and the form, to map the infinite with the finite.

Take my own design of Prisoner of war (POW) museum of Jianchuan Museum Complex as an example, the form was consciously weakened in the architectural creation. What I want to highlight is an atmosphere and mood – depression, distortion, pathos… This is my expression of the psychology of the special group of prisoners of war, and in this way to impress the audience and strengthen the artistic impact of the building.

It should be said that this understanding of beauty is also reflected in the works of some contemporary western artists and architects. For example, in the oil paintings of Italian painter Giorgio Morandi, through those repeatedly arranged bottles with simple shapes, I can also feel a similar pursuit of transcending objects and achieving a certain artistic conception. In the architectural works of modernist architect Luis Barragán, we can also feel the beauty of this "conception" in those quiet geometric spaces. It is only in the western context that there is a lack of self-consciousness about such an aesthetic experience. In the west, artistic conception has never been an independent aesthetic category, which is why there is no word corresponding to "artistic conception" in English. However, in Chinese aesthetics, the pursuit of "artistic conception" has always been the proper meaning of artistic creation. I hope that contemporary Chinese people can learn to appreciate the beauty of this "conception" again, and I also hope that Chinese architects can express the beauty of artistic conception more consciously and fully, which can also be said to be the natural advantage of Chinese architects.

Fei: Through your previous explanations, I have an understanding of "intellectual state" and "artistic conception". In your theory of architecture, intellectual state, artistic conception, and language are together a system. As you said before, language is the carrier, not the essence. This is very different from western philosophy. Could you please expand on this point and talk more about language?

Cheng: In the West, the 20th century was the era of philosophy of language. Heidegger said that "language is the house of being" and Derrida said that "there is nothing outside the text (Il n'y a pas de hors-texte) ". The prevalence of digital language in recent years has established the dominance of linguistic philosophy in the West. Reflecting this, architecture is now a world of languages. From the pattern languages of a few decades ago to the parametric ones of today, the influence is extensive.

This is how I see it. "Language" contains semantic meaning, especially it is worthy of recognition for its rational analysis of the creative mechanism of "perception but not speech". However, at the same time, we should also see that it ignores the deep connection between all things and people's cultural and psychological emotions, which makes it difficult to fully explain and reflect the reality of architectural creation. It is for these aspects that I propose such a trinity theoretical system of intellectual state, artistic conception and language.

In my opinion, if architectural creation follows the western idea of language as an ontology, it is very likely to go astray into the formalism of emphasizing "external images". Since the second half of the 20th century, the philosophical cognition of language as the ontology combined with the post-industrial civilization, western culture has gradually shifted from the pursuit of "originality" to the pursuit of "pictorializing" and "spectacle". In the field of architectural design, there is a tendency to pursue the sensory stimulation of architectural forms. Faced with this phenomenon, should we also think about whether this idea of taking language as the philosophical ontology and focusing on external forms has its limitations?

Therefore, I think it is time to reflect on the western way of thinking that always goes in circles on "language" and "form". In contrast to the west's special emphasis on form as a philosophical ontology, we find that the Chinese rarely talk about language and form in isolation. Whether it is "writing the spirit with form" or "words to express" or "words to reason", language and form are combined with the emotion and meaning they are meant to convey. This illustrates the Chinese understanding of the relationship between "form," "concept (artistic conception) ", and "reason (intellectual state) ". Will such a viewpoint shed light on contemporary Chinese philosophy, aesthetic construction, and the development of contemporary Chinese architectural theory?

Fei: Can you tell us specifically how you think about the relationship between "language", "artistic conception" and "intellectual conception" in your architectural creation?

Cheng: First of all, as already mentioned, language cannot exist separately from the artistic conception and intellectual state it expresses. The ultimate goal of language and form in architectural creation is to create an artistic conception and express the intellectual state. Once language is stripped away from consideration, it loses its creativity and vitality, or becomes a "spectacle" architecture purely for the purpose of stimulating people's senses, or like the once popular "international style" architecture, falls into rigidity and repetition.

Secondly, there is no priority between the three, nor is there a relationship of who decides who. On the one hand, it is true that "art is the way" and "speech is the expression", but on the other hand, it is also true that "the way and art help each other grow", and they are one with each other. Language and form innovation cannot be achieved without the support of philosophy and aesthetics, but if language is missing, architecture becomes an abstract idea with less heat of life and emotion.

Thirdly, since form is the carrier, the architects can choose various means to better express their ideas in the creation. In particular, compared with the relative stability of "meaning" and "reason", "language" will keep changing with the development of the times. Architects need to transform and innovate on the basis of fully mastering the language of ancient and modern architecture in China and abroad. Therefore, I strongly disagree with the practice of using a collage of elements in architectural design, or simply expressing the spirit of Chinese culture in a certain style, such as the current popular New Chinese style. Each project has different conditions, so how can we use the same language to express it?

A while ago, many people were talking about "strange" architecture. I think we have to understand the criticism of the General Secretary Xi on "strange" architecture, and we cannot use it to negate architectural innovation. I think we can't negate architectural innovation by opposing "strange and bizarre" architecture. Architectural innovation is to "keep the right and make the strange". The key is to be "strange" with "feeling" and "reason". Take my just completed Wenling Museum as an example, this building uses a non-linear formal language to shape the form of a mountain rock. Some people may find it a bit "strange" when they look at it. But it is precisely because of such a choice of formal language that the museum is able to create its own aura in the midst of the chaos of the surrounding city, and at the same time echoes the local stone culture. So, if we look at the building in the context of its urban environment and culture, we will not find it "strange" at all. It is also because of the cultural heritage behind this architectural form that it is distinguished from some programmed non-linear forms and has a Chinese flavor.

As I mentioned earlier, the theory of architectural creation, which is composed of "intellectual state, artistic conception, and language", is my own opinion based on traditional Chinese philosophy and aesthetics, from my own practice and thinking, and facing complex realities. Although it may not be mature, it is useful not only for the current reality, but also for the development of architectural disciplines and even cultural and scientific fields. Chinese architectural theory research urgently needs to get rid of the shackles of the western "classics" and find answers from our cultural background and our own practice on the basis of equal cross-cultural dialogue. And similar discussions and explorations should be more and more daring. Only in this way can Chinese architectural design truly gain vitality and freedom.

Fei: Generally speaking, not many practicing architects pay attention to theoretical research, let alone form such systematic and complete theoretical ideas. So, in the end, I would like to ask you to tell me how you, as a practicing architect, build your theoretical thinking structure step by step?

Cheng: Much of my thinking about architecture comes from my own architectural practice journey. It did not start with a framework, but gradually developed a relatively complete understanding over time. Several of these projects have had a great impact on the growth of my personal thinking.

Take the Huanglong Hotel designed in the early 1980s as an example, the holistic thinking in traditional Chinese philosophy inspired me greatly when our design team won the international bidding. Through this project, I began to think seriously about how to translate traditional Chinese philosophical and aesthetic ideas into architectural design. During this period, the relationship between eastern and western cultures was a question I kept thinking about. From the mid-1980's "Foothold of Here, Now and Self"[9], to later paper *Reflections between History and Future*[2], and 10 years later in paper *Rethinking between History and the Future*[10], I have a clearer understanding of how my path of architectural creation should go.

During this period, I encountered different types of architectural design projects such as the National Theatre of Ghana, the new Hangzhou Railway Station, the Master Hongyi Memorial Hall, the Prisoner of War Museum of Jianchuan Museum Complex, and the Zhejiang Art Museum. During the creation process, I have deeply experienced that a good architectural proposal needs to deal with the relationship between various elements such as function, form, site, technology, economy and culture in a comprehensive way. This is a process in which the rational and irrational are constantly compounded and transformed into each other, and architectural design is to find the right balance point. This caused me to think about the mechanism of architectural creation and deepened my understanding of the spirit of Chinese culture. In the early 2000s, I proposed a more complete, meso-level concept of "unity of heaven and man," "unity of reason and image," and "unity of emotion and scenery" from three levels: epistemology, methodology, and aesthetic ideal. I put forward a relatively complete theory of architectural creation at the meso level. After 2010, I began to think about whether I could start from the philosophical and aesthetic level, connect the ancient and the modern, integrate the east and the west, and establish an architectural theory system based on the contemporary Chinese context, which is the three levels of intellectual state, artistic conception and language. Therefore, my theory is not a framework or system in advance, but a summary of the project to think about the direction, the philosophical aesthetics and the method. While the ideas are gradually sprouting and forming, I go to books to find a kind of proof and resonance, and then intertwine to form my own thought framework.

I have always believed that "thinking" in the field of architectural design is not abstract metaphysical thinking, but should be combined with "doing", which is thinking in practice and practice in thinking. Therefore, the relationship between theory and practice is not simply theory guiding practice, but "unity of knowledge and practice". I hope that these ideas and theories I put forward can be discussed by architecture scholars and architects, and I also hope that more people can care about these researches and devote themselves to them. The discipline of architecture is ambiguous and uncertain. There is no standard answer, no golden rule, so there is a lot of room for exploration, and this is where the charm lies. I am still optimistic about the future of architecture, including the future of Chinese architecture.

NOTES

① See Xie He, *the Record of the Classification of Old Painters*.
② See Liu Yuxi Dong's *Wuling Collection*.

REFERENCES

[1] Liu Xianjue. Modern Theories of Architecture. Beijing: China Architecture & Building Press, 2001
[2] Cheng Taining. Reflections between History and Future. Architectural Journal, 1989(2): 39-41
[3] Yue Daiyun, Alain Le Pichon. Dialogue Transcultural 4. Shanghai: Shanghai Literature & Art Publishing House, 2000
[4] Xu Wenyu, Lectures on Zhong Rong's Poetry, Lectures on Poetic Remarks on the Human World. Chengdu: Chengdu Ancient Books Publishing House
[5] David Smith Capon. Architectural Theory. Beijing: China Architecture & Building Press, 2007
[6] Albert Einstein. The God Letter, Einstein Collection. Beijing: the Commercial Press, 2010
[7] Cheng Taining. Cheng Taining Architectural Works: 1997-2000. Beijing: China Architecture & Building Press, 2001
[8] Cheng Taining. My Philosophy of Architecture, Contemporary Chinese Architects (Vol. 1). Tianjin: Tianjin Science and Technology Press, 1988
[9] Cheng Taining. Foothold of Here, Now and Self. Architectural Journal, 1986(04): 11-14
[10] Cheng Taining. Facing the Future, Going Your Own Way – Rethinking Between History and the Future. Architectural Journal, 1997(1): 7-10

作品年表（2015-2021）
CHRONOLOGICAL LIST OF PROJECTS (2015-2020)

2015　福州市第一医院外科大楼
中国·福建·福州
完成方案至初步设计工作
52958 m²

秦皇岛七里海国家港湾会议中心方案
中国·河北·秦皇岛
未落实
30000 m²

郑州博物馆新馆方案
中国·河南·郑州
未落实
62000 m²

天津美术学院方案
中国·天津
国际设计竞赛未中标
200000 m²

龙岩市委党校（126 页）
中国·福建·龙岩
已建成
98347 m²

厦门同安新城（丙洲片区）
中国·福建·厦门
中标，作为总控单位设计深化，建设中
1600000 m²

2016　南京美术馆（14 页）
中国·江苏·南京
国际招标中标，基本建成
97275.9 m²

青岛（红岛）铁路客站（138 页）
中国·山东·青岛
2016 年设计，2020 年竣工
407380 m²

南京城墙博物馆改扩建工程概念方案设计
中国·江苏·南京
未中标
10000 m²

苏州中学苏州湾校区（158 页）
中国·江苏·苏州
2021 年竣工
90307 m²

宛平剧院改扩建工程
中国·上海
未中标
27000 m²

2017　杭州萧山国际机场新建航站楼方案设计及陆侧核心区总体规划概念方案
中国·浙江·杭州
国际竞标入围
551260 m²

首钢世界侨商创新中心方案
中国·北京
未落实
294685 m²

中华国乐中心·江阴（268 页）
中国·江苏·江阴
施工图设计中
108340 m²

北京城市副中心大剧院建筑设计方案（254 页）
中国·北京
国际竞标优胜方案
121200 m²

杭州西站枢纽核心区城市设计暨建筑概念规划方案
中国·浙江·杭州
竞标中标
500000 m²

长春世界雕塑艺术博物馆（70 页）
中国·吉林·长春
2018 年竣工
18000 m²

西安交通大学新校区博物馆及多功能阅览中心（104 页）
中国·陕西·西安
2019 年竣工
35566 m²

2018　南京熊猫中山东路 301 地块 2 方案
中国·江苏·南京
竞标方案第一名，未落实
200000 m²

嘉兴市委党校迁建及学苑广场方案
中国·浙江·嘉兴
未中标
250000 m²

江北新区图书馆
中国·江苏·南京
施工中
77272 m²

龙岩闽西宾馆迁建工程
中国·福建·龙岩
完成方案设计
50000 m²

南京永济寺地块设计方案
中国·江苏·南京
未落实
11729 m²

2019　中国第二历史档案馆新馆方案
中国·江苏·南京
竞标方案入围
88752 m²

厦门一场两馆、新会展中心城市设计及
建筑概念设计方案（202 页）
中国·福建·厦门
国际竞标中标
6366200 m²

厦门新会展中心（210 页）
中国·福建·厦门
施工中
1064000 m²

衡阳市博物馆和华侨城艺术博物馆
中国·湖南·衡阳
设计中
39000 m²

中国水工科技馆（304 页）
中国·江苏·淮安
施工图深化中
107500 m²

南京雨花华邑酒店（278 页）
中国·江苏·南京
施工中
45627 m²

东南大学九龙湖校区工科综合科研大楼
中国·江苏·南京
竞标中标，设计中
146826 m²

南京北站
中国·江苏·南京
完成概念方案设计
400000 m²

杭州西站及城市综合体（184 页）
中国·浙江·杭州
施工中
510000 m²

易俗历史文化街区更新改造工程概念设计
中国·陕西·西安
完成概念方案设计
41000 m²

北京新国展二、三期城市设计及建筑概念
设计方案（226 页）
中国·北京
国际竞赛优胜方案
813575 m²

2020　徐州园博会宕口酒店（290 页）
中国·江苏·徐州
施工中
29597 m²

国深博物馆建筑设计方案（244 页）
中国·广东·深圳
国际竞赛未中标
127898 m²

深圳改革开放展览馆方案
中国·广东·深圳
未落实
90650 m²

雄安新区启动区 05、07 地块创新坊标志建筑
中国·河北
概念建筑方案设计中
187238 m²

2021　苏州大剧院规划及建筑概念设计
中国·江苏·苏州
国际竞标
30000 m²

杭州市轨道交通 TDO 综合开发地铁海潮站城
市设计
中国·浙江·杭州
国际竞标中标
599364 m²

南京溧水文化艺术中心概念设计
中国·江苏·南京
完成概念方案设计
45000 m²

雄安新区启动区创新坊标志建筑概念设计
中国·河北·雄安
完成概念方案设计
203453 m²

嘉兴平湖市明湖国际科创岛城市设计
中国·浙江·嘉兴
国际竞标中标，设计中
560000 m²

后记
POSTSCRIPT

一、本书与中国建筑工业出版社出版的《中国建筑师——程泰宁》《程泰宁建筑作品选 1997-2000》《程泰宁建筑作品选 2001-2004》《程泰宁建筑作品选 2005-2008》《程泰宁建筑作品选 2009-2014》《程泰宁建筑作品选 2015-2021》相衔接，并保持体例上的一致。

二、本书的版式设计由王琼宇同志负责；照片部分由陈畅同志负责。

三、我的学生参与了本书的出版工作：
黄卿云负责本书的英文翻译与校审工作，刘鹤群参与了部分校审工作。
曾媛、林子苏为本书部分效果图做了后期调整工作。
由邱培昕牵头，李尚媛、汪昭涵、陈一凡、武诗葭、彭舒妍、张廷昊、邱健雨、蔡万成、韩佳琪等同学绘制了技术图纸。

四、中国建筑工业出版社对本书的出版给予了大力支持。

谨对以上同志表示诚挚的感谢。

图书在版编目（CIP）数据

程泰宁建筑作品选=CHENG TAINING ARCHITECTURE WORKS 2015-2021.2015-2021 / 程泰宁著. -- 北京：中国建筑工业出版社，2020.12
ISBN 978-7-112-25469-9

Ⅰ.①程… Ⅱ.①程… Ⅲ.①建筑设计－作品集－中国－2015-2021 Ⅳ.①TU206

中国版本图书馆CIP数据核字(2020)第180188号

责任编辑：徐明怡　徐　纺
责任校对：王　烨

程泰宁建筑作品选 2015–2021
CHENG TAINING ARCHITECTURE WORKS 2015-2021

程泰宁　著
*
中国建筑工业出版社出版、发行（北京海淀三里河路9号）
各地新华书店、建筑书店经销
北京雅昌艺术印刷有限公司印刷
*
开本：889毫米×1194毫米　1/12　印张：29　字数：679千字
2022年11月第一版　　2022年11月第一次印刷
定价：395.00元
ISBN 978-7-112-25469-9
　　　(36459)

版权所有　翻印必究
如有印装质量问题，可寄本社图书出版中心退换
（邮政编码 100037）